Robotics

Discover the Robotic Innovations
of the Future

*(A Beginners and Advanced Guide in
Understanding Robotics)*

Lanny Hansen

Published By **Ryan Princeton**

Lanny Hansen

All Rights Reserved

Robotics: Discover the Robotic Innovations of the Future (A Beginners and Advanced Guide in Understanding Robotics)

ISBN 978-1-77485-640-6

Legal & Disclaimer

The information contained in this ebook is not designed to replace or take the place of any form of medicine or professional medical advice. The information in this ebook has been provided for educational & entertainment purposes only.

The information contained in this book has been compiled from sources deemed reliable, and it is accurate to the best of the Author's knowledge; however, the Author cannot guarantee its accuracy and validity and cannot be held liable for any errors or omissions. Changes are periodically made to this book. You must consult your doctor or get professional medical advice before using any of the suggested remedies, techniques, or information in this book.

Upon using the information contained in this book, you agree to hold harmless the Author from and against any damages, costs, and expenses, including any legal fees potentially resulting from the application of any of the information provided by this guide. This disclaimer applies to any damages or injury caused by the use and application, whether directly or

Table Of Contents

Chapter 1: What exactly is robotics, and how can it be applied to your life?

Robotics is any type of non-human controlled machine, modernized.

Automates are basically conscious. This allows us to expand this idea further by saying that robots have to be capable of thinking and acting as humans. A robot of human intelligence could make decisions, plan actions and take them to completion.

How do robots operate?

The technology is complex as the machinery. However, the human brain as well as the human nervous systems are extremely complex. They can only perform complex intelligent functions.

Robots build their robots by using the human mind rather than copying it. Each robotic system has its own function. Each algorithm draws on the human mind with a variety of algorithms that are based upon auditory, visual, and olfactory sensory systems.

These algorithms were created to match the specific situation in which the robot is used.

Robot programming often uses the following:

Software: Programs written in programming language. Every programming language has its own set of rules. To code a VCR remotely, for example you might use specific rules.

Robot hardware refers to the robotic body or platform as well electronics. The robot's brain is typically complex and robust and includes smaller, lighter, or more economical, rubber or plastic units. Robot hardware:

Comportemental software algorithms: These software algorithms detect the physical conditions of the robot, the input it gets and the interpretation of that decision.

Cognitive algorithms. These algorithms decide what the robot should perform based on past actions and environmental knowledge.

These algorithms use past decisions and knowledge from the environment to determine what the robot is expected to do. These are physical algorithms, which combine the two prior ones.

It is possible to design a robot program capable of creating its own rules. Data processing algorithms : These algorithms enable a program, through sensors and software algorithm, to process large amounts of data. This allows them to create plans.

These algorithms enable program software to sort out the vast amounts of data generated by the sensors and software software in order build a plan. Artificial Intelligence is an artificial intelligence system that mimics the human brain. It allows users to think about their own thoughts.

A robotic example

The human body can take many decisions on its own. It is an extraordinarily complex engineering part. We make millions of choices every day without ever realizing it. These are the behaviors that a robot can learn by looking at its environment and comparing what it sees with what it learns.

Here's an example for a robotics system:

The robot is able to analyze the forces being applied by each move of the arms and decide what actions it should take against them.

The robot then executes these actions:

To counter the wall's pressure, he pushes the robot arm back.

Then it places its arm under its arm, so that it is no longer pressing against the wall.

Then, it pushes a block against the wall.

These are the small parts that allow for movement.

Imagine that these few actions make programming them very easy. It's not as simple as that. The human brain is capable of only one action at once. A person cannot learn to run and walk at the same time.

The robot should follow a set instructions so that it is not able to create its own algorithms.

When a robot encounters something he doesn't get, he must be capable of "thinking" about himself through analysis of all the information.

It is not good to have all the software algorithms and sensors available, but it is not possible for the robot interpret the data.

Biosecurity algorithms

A neural networks is a form of artificial Intelligence that replicates the brain function. Building robots involves copying the brain.

Neurons have been programmed to predict how they will react. A neuron is programmed to produce the correct output when the output of an electronic screen changes.

In order to make a robotic system with this set-up, it is important to create a robot to answer specific commands. One example is a robotic arms. The robot arm has a number metal rods. A single rod includes the electrodes.

This is the job of the robot arm, which interprets and responds.

A new robotic system was developed called myoelectric propthesis. This system uses electric signals to control a robotic hand. They are transmitted via the headset using electrodes to the wearer of the headset. These electrodes send the position and move of the hands to a robotic arms.

With neural networks, similar principles can be applied. The idea is to build a system that can draw upon the behavior. This means the robot should not need to be taught how it works.

It's exactly the same way deep learning neural networks can learn about a face and an animal from images. They only need to look at images that have been provided by many people.

Additionally, it will be necessary to create stronger systems for creating life-like robots.

The ability to recognize and analyse the information received by robots is impressive. As humans, however, we can make terrible decisions based only on a handful of pieces of information. We are wired to feel emotions. The more information we have the better.

To create truly lifelike robots, artificial neural networks must be developed that can simultaneously perceive and interpret many data.

What is robotics all about?

Answer: "What's robotics?"

Follow this discussion. If you have any questions, e-mail us.

It is against the law to post any advertisements, profanity or personal insults. For more information, refer to the CNET Forums guidelines.

If you're looking for technical support, ensure that all your system information is included, including your operating system and model code. You are encouraged to post your opinions in the forums.

How do I become a robotics engineering engineer?

Want to be a robot designer? Robotics kits and programming make robotics and programming more accessible and affordable than ever.

The easier it is for children and their parents to program the robots, the cheaper they will be. There are many Lego Mindstorms products that can be used by children interested in robotics.

But, let's be careful: even though you may not have the huge financial resources of a corporation, don't worry. There are plenty of other large LEGO kits that can offer lots more entertainment.

First, the Lego Mindstorms EV3. The robot development kits have been replaced but still pack a huge punch. The NXTSTEAM platform for building and programming lego robots is now available.

The kit includes ten NXT Mindstorm 3 blocks, a LEGO NXT Creator "computer", Wi-Fi adapter and a LEGO E V3 Wheel, as well as a Bluetooth sensor. Laser and an engine.

Five configurable control sliders and a Cortax M4 MCU 1.6GHz ARM, GPS, codec, micro SD card, and a 6-hour running battery pack are included with the NXTNXT Creator Computer.

The EV3 consists three major components.

Connectivity -- The EV3 is equipped with four Bluetooth and Ethernet wireless sensors. They include an infrared search sensor, a temple-altitude barometer sensor, a location magnetometer and a compass.

The all terrain and force/torque block comes equipped with two rubber tires and two metal wheels. This should make moving buckles much simpler. The inclined and rotary drums on the four-axis are equipped with four metal-and-plastic covers. Four servo motors, which rotate according to how far you tilt the drum, have motors.

How to learn robots?

Google's Application will show your how to

Robotics is a skill that many of the most successful roboticists around the globe want to master. They want to show you why a machine does the things it does. Marc Raibert (Boston Dynamics co-founder) started working on something simple. An app that is free and easy to use allows you to control a humanoid-robot device.

How to do robotics

Robot technology is how we educate our students about interdisciplinarity at the Middlebury Institute for International Studies. Robots can help us invent new technologies and promote our other instruments, which are very difficult to do manually.

With robotics becoming an established tool in science and technology, they are also a popular way for people to interact and learn from machines.

I learned a lot as a software developer, particularly programming in C++. Many of our robots will require large quantities of manually-written programs, including task runners, toolskits, payloads, as well as toolkits. Our most important skill is to be able to comprehend algorithms and other components of software like task runners, payloads, and toolkits.

Technology challenges

The way that we study the natural environments has been transformed by robots. They use them in the analysis of the earth's surface, and when using aerial cameras to study wildfires. Drones can be used to collect exceptional quality data by researchers to help better understand wild animals.

We are just starting to build a range robotics that can interact and collaborate with us. Robots such as humanoid bots can interact directly with us, allowing us to move them around in the laboratory, provide instructions and tools that will help them complete their projects.

Many companies place emphasis on humanoid machines when creating robots that interact with wild animals. These robots have the ability to respond and acknowledge people. They can also perform simple gestures with their hands. These robots look awkward, but it remains to be determined if they are able to function in the same environment that humans.

Robots have been used in Iraq, Afghanistan, and elsewhere to detect improvised bomb devices and assist with bomb removal. Groups are currently working to create robotics that can more effectively search explosives, such by identifying their signature and dedds.

Machines versus people

Even though robotics can make life easier, we have to be aware that it will always be in competition with human subjects in certain ways. People still want to work on robotics projects, and not because it is required. A robotics engineer is more likely than anyone to encounter a problem when his robot is outside the city and cannot return to solve it.

Robots are steadily improving, and the question now is: Can they ever catch up with human intelligence? While it's unlikely this will happen soon, robotics continues to grow and develop, so it is safe for us to bet that robotics will one day take over all aspects of human societies.

What is the history of the first robotics system?

Are you ready to be part of the IFLScience Handbook to The Robot Revolution? We will be looking at all types of robots: from the simple, low-tech rovers that you can hold in your palm to the complex wooden machines that could one day end humanity. Robot products have the potential for solving all kinds of problems. These include how to feed an ever-growing global population and clean up our space.

If it sounds interesting, you can scroll down to find out more about the best bots.

Robots are in many ways the reason that humans exist today. They were critical for our wars and exploration, and they will be crucial in the healing many of humanity's problems.

The robot we often think of is large and versatile. Particularly robots intended for industrial uses can accomplish many tasks at different sizes. Their only limitations include their weight, size, and cost.

Spiral Jet engine 2nd-class Marine Attack Platform for Ship-to Shore

This 22-foot high ship-toshore aerial platform had many forms in the last century. These ranged from doughnuted ships for ship patrols and soldiers to flatbed trucks-like "shiptoshore", which saw action during World War 2.

It's equipped with two Rols Royce Merlin Twin 1,000hp engines, a separate diesel-powered engine and a hydraulic spring. They are all mounted onto two heavy-steel roller composite tracks. This combination makes it one among the fastest ground-mounted vehicles.

Robot King Crab Mining

You will most likely end with something like it if a mobile mechanical Scorpion is built. It was designed to find hidden minerals in areas that are unstable and make this car very efficient, compact, and strong.

King Crab mines on salt-rich, undercarried flats in North Iceland (and Southern Canada). Each trip can carry up 100 tons of material. This is more than most miners have the ability to carry. King Crab also features a lightweight construction to allow it to move swiftly and safely through the mine.

The vehicle uses so little energy it can mine until it has to be refueled for three years. Only a few mechanical vibrations can stop it from mining.

The king crawler is a single turbo diesel vehicle with three 32-hour diesel engine that provide power. It has a 72 percent heat efficiency maximum. This means that energy will be used for motors, rather than being lost.

Exobots can be one of most impressive robotic systems around. They are a system that you can control via your smartphone. These autonomous robots are capable of extracting and collecting natural resources such water, metals, and biological material.

It's not only an investigative robot but one of the most useful. Remotely check Exobots and send them to the person to help diagnose or collect specimens. Exobots were also used as UAVs to aid in rescue and search efforts.

At the moment, approximately 50 robots worldwide are in use. A number of companies have produced new models or been working on improving existing models, including AeroVironment Megascale Robots (Megascale Robots) and Swarm Dynamics.

Happy Luddite

Luddites - 19th century engineers and inventors - considered robots a threat because they thought robots could not match the work of human workers. They believed robots' minds and bodies were superior to humans, and that human-like robots would only lead to more suffering.

"Luddism," an expression of opposition towards technological change in England was named after Scottish workers who had broken machinery, causing a lot of industry to collapse in the region. But what would luddites have to do if they were given robots for destruction?

There have been several films that depict the use of robots in combat. Many real-life robots, however, are being developed with the goal of supporting military applications. The military has considered drone swarms as an option to humanoid and humanoid-robots.

Stripettes

Mack Trucks is collaborating with Daimler Trucks to create the Stripebot over many years. The Stripebot has an 8 meter robot arm and is long articulated. The original idea behind the Stripebot is to manipulate objects at different heights and angles.

The stripebot has the ability to work with humans in safety, can perform repetitive tasks, is able to help build teams of robots, and can even be used for group projects. The robot arm is strong enough lift heavy objects. It also allows for easy dexterity, flexibility and mobility.

ZetaLab & iRobot

iRobot and Cornell University teamed-up to create a tie robot fish, which could be used for underwater, marine explorations and environmental inspections. Colin Angle was the president and CEO of Colin Angle. He wanted a simple robot that could inspect under water, even in potentially dangerous environments.

The fish was, however, also astonishingly capable. It was equipped with a camera for photos and a sonar sensor to detect submarine objects. This combination of abilities made it well-suited to underwater tasks. iRobot released him to the public in May 2015.

Universal Soldier portrays Nicolae Carpathia's role as a luddite that wants to destroy humankind. Carpathia appears to be an android controlled from Martin Sheen. He hides his true nature from the general public. It wants to release an army remotely controlled robot army and has military bots in the Universa collections

X is Google

Google X is a division at Google that develops and investigates technology that can improve the quality its products, services, platforms and other offerings. It is located Mounta View and focuses primarily on "moonshot", projects that fall outside of the traditional fields of opportunities.

The car was modified from traditional vehicles. It uses digital maps to navigate and has no steering or pedals.

What are the 3 laws of winter robots?

(Physics World). Thursday, 10 January 2019, 1:00 UTC. Winter Robotics 2019 officially began. It saw around 360 people attend the Berlin Elbe Academy. Science Foundation broadcasts its first live stream via YouTube.

The Winter Robotic Conference in Dublin will be co-organized with colleagues at the University of Tubingen, Germany by Science Foundation, the Irish Research Center for Robotics based at Trinity College Dublin.

The agenda includes the Winter Robotic Research Forum. It is a three-hour conference that features presentations by people from around the globe. Additionally, there are a number of other lectures and workshops scheduled throughout the conference.

This year's theme was "RoQ - Robots to reskill."

Chapter 2: What is the name of the first robot that we made?

David Hume (Scottish Enlightenment philosopher) famously noted that a hundred men could design, construct and maintain an autonomous robot capable of functioning indefinitely without the aid of a human monitor 250 years ago. This idea has been taken up by many in the robot industry. (Quarto Group)

There are now many automated and intelligent "smart work robots" systems that can do a variety of tasks. Other smart machines include the Roomba vacuum, which is a simple smart machine for everyday use. Baxter, another leader, is known for their innovative auto-programming which allows them to do far more than just vacuum the product.

What happens when these systems become integrated with intelligent machines?

Imagine an increasingly automated factory, or any other building that can be accessed by a variety of smart devices. What are the benefits of thinking about the wiring that connects buildings to the rest the world?

What can you do to make a difference?

A smart workshop is one where the robots can work independently. This allows us to spot potential problems and allocate the resources accordingly.

If there's a problem we need to fix, we can send a message to the robots to tell them. It is possible to program robots in order for them to be notified of any problems and we can then take appropriate action before they cause further damage.

Robots would then be able to interpret human inputs, such as noises and vibrations. While safety issues are critical, many building owners might have known this already.

Are robots capable of learning at work?

(University of Applied Sciences Stuttgart - What if a machine is loosened inside a plant and learns new jobs?

These are the central questions in a research effort led by Georg Feichtner. Georg Feichtner is also a professor of robotics at Stuttgart University and the head of the university's research team.

The researchers decided to equip the car with a learning program that could analyze a large range of sensory data.

The prototype vehicle for the research team was equipped with sensors for detecting people and an additional sensor to measure and analyze the environment input, and then decide what to make.

Researchers used the vehicle to complete a range of tasks to investigate potential uses for robot-driven vehicles in mobile sensors.

The vehicle was tested in a first inspection.

The robot could make use of the scan data and vision to find the most efficient route to passengers.

Next, researchers used remote control technology to transport the vehicle over a pedestrian crossing. The goal was to detect pedestrians before they cross the road.

Finally, the vehicles would have to be transported to a construction site. It would be one of the first vehicles to go on a site where many sensors are used to identify and classify different sensor types.

The vehicle was capable and competent in performing the task. This feat was achieved despite the fact that they did not know about the site.

Is it possible to teach a robot how to collect waste

The purpose of this project was to discover what it takes in order for a robot to learn how to perform a given task.

Feichtner's team was faced with a variety of tasks. Feichtner wondered if the robot could be trained to sense objects in its environment before it has seen them.

In the second stage, the robot was tested for its ability to detect specific items (e.g. a bottle of mug, bag) and could then be used to classify objects by looking at information both from the object as well as the environment.

The team reported in both tests that they did it in different ways.

The robot's sensors picked up the object from the "Tash Bin" in the first example. However, the robot system was asked to identify the objects as a "bottle", cup, or bag. The robot had not seen the object or been trained in this way before.

In the second instance, the researcher was required to tell the prototype which objects were previously seen and which ones had not.

Further tests are planned to verify that the robot is capable of learning when and how to properly classify objects.

Future projects will explore whether the robot could perform these tasks by itself, as well as observe and learn about how humans interact and interact with robots in this environment to increase its ability to work efficiently.

Copyright KAIST Image title Can a robot learn the art of collecting waste?

Copyright

BBC News spoke with Prof. Feichtner, who stated that "The goal of creating a robot can handle multiple tasks." Recent years have seen many tasks being identified, including self-driving car, but each task is unique.

"For example, there is a variety of parameters for driving and handling a mug of tea on a winding roadway. But, we discovered that robot-learning technology has advanced to a point where we can begin to develop a generic system, such a robot, that can handle a multitude of tasks in a similar environment and in the very same situation.

"We are doing it, and there's some companies that deal with similar problems." Lockheed Martin makes one. We also have research partnerships, including with US companies like Google and Amazon.

To what purpose are robots used?

This is the perfect object for testing.

Welcome to Wasp. Wasp serves as an interactive classroom/office in support Ubiquitous Robotics education.

In this case automation has helped maximize efficiency. Robotics has become a weapon. In the future, robots could become an integral part to our everyday lives.

Would you like the opportunity to apply for robotics projects?

We are a group consisting of NIT JEE 10/11 students and professional programmers who love Ubiquitous Robotics. We aim to increase Ubiquitous Robotics. To achieve this goal we aim to promote Ubiquitous Robotics, and Ubiquitous Education with innovative resources.

What salary does a robotics engineer make?

This wage calculator is available

Recently, The Higher Education Chronicle requested that I analyze salaries for various types of college leaders.

After speaking with you about the big picture, we realized that there's more to learn than the intellectual brilliance a Nobel Prize-winner has. How much is the role worth? How long will it last? What interpersonal dynamics do you have to work with in order to make the company successful? But that's just the beginning.

Here are some ways to understand the data.

It is helpful to compare the number and salaries of full-time faculty with teachers at the beginning. Teaching ten times the number of students you can as a full professor is more than what the average college president can do (and many colleges cannot afford to hire even one person in their faculties).

Additionally, full-time students have access to more faculties at some institutions than they do full-time, while full-time students have access to fewer faculties at other institutions. It is possible for a university to claim that the president is a tenured faculty member, but the difference is not that large.

You'll also notice a pattern if the faculty pays you a full time salary. Salaries are higher for Presidents and Vice-Presidents of large urban colleges with lots of students. However, these salaries can also be very high if the college is large enough to attract people from all over the country.

The University of Wisconsin-Madison President's Salary is $427.369. That amounts to $148.717 per Year. This same data shows that the Florida State University president's salary is US$476,004, a sum equivalent to US$144,000. I find that this is not unreasonable.

The University of Wisconsin-Madison presidents get a salary of 16 faculty members. It is not surprising that this is so high. But, if you consider the fact that the president gets 12 faculty member salaries if we split the salary of his/her salary into the number of full-time faculty.

Take into account that presidents are more highly-sought-after scholars and Nobel winners than they are in terms of their salaries. Because the American system is based on a small percentage vote, the president is often a superstar who makes more money.

Even though you may believe that the president should have a large salary, data clearly shows that even if there isn't much open space, there is no reason to expect a president to have a high income.

Inequality in pay between professors, and presidents. Recent wage data I found on college presidents' earnings shows that there are significant differences in professors' pay and college heads'.

According to data, the President if a university has fewer than 5K students is paid around

$185,000 annually while the President if a university has more than 25K students receives around $306,000 annually.

There is a wide variation in the ratio between the President's salary and that of the Professor. Ray Watts, who is president of University of Alabama Birmingham earns $399,000 each year, while Ray Hawkins, an apropos university of Alabama, earns $187,000 per year.

The ratio between president's wage and faculty's wage is what we need to be focusing on. This ratio is much higher if you compare wages to professors. It is about 70x higher than the salaries of faculty members at my university. This is the only university where students can't attend.

There could be large discrepancies. If the president gets paid $600,000.00 each year while the person who runs book fairs receives $40,000 annually, then we have to ask, "Is that fair?" There may be huge differences in the cash distribution at university. My university has many high-quality professors.

The president's salaries are not much higher, but they are still lower than enrolment. It is difficult to compare salaries between faculty and university students. The UC presidents have about 150 times the faculty.

It's possible to think that a chairperson should make less than the average professor if they have a doctorate. I believe this is perfectly fair. It is a culture inwhich only doctors and scientists are treated with respect. But that's not the case in many universities. There are many people who have multiple degrees at most universities.

Indeed, I think it's great that the president receives high-paying salaries. The institution wouldn't be able to give a clear picture if it didn't pay the president well. However, the president is a prominent figure who represents the university.

Faculty members who are unhappy with their salaries are fine. But my concern is that President of University of Alabama has made 70% more than our professorship. For me, this is unacceptable. While I do not claim that what is actually worth it is not paid,

Chapter 3: How to make Lego robotics

Not use official instructions for building blocks

Lego is trying to revolutionize robotics. This has attracted increasing interest in engineering and science.

However, robots can now also be constructed using the same bricks as in the 1980s and 1970s.

How do I make my own robot? How do you do it? Where should you start? And most importantly, why are you so interested in your robot's capabilities?

Remember that there are many options for building a Lego robot.

Lego robotics has a lot to offer. The set of instructions, bricks and simplicity are part of its appeal. But robot rules do not exist. Robot builders can create and modify their robots in many ways, including software programming.

You can build the equipment you need

The first part will be a simple robot.

The robot's main body can be seen in the above block. There are several types of robots that you can build, e.g. Make the legs and then transform your body into a larger, smaller robot.

However, to create a robot that is able to stand on its two legs, it will be necessary to make two or three basic blocks with the exact same basic shapes, provided they both have the right shape factor and the number of bricks.

Programming an algorithm for one robot allows us to program it in a way that will determine when it is appropriate to complete a specific step and then continue with the planned behavior. Divide and conquest is a technique to build machines that are simple yet efficient. They can also function autonomously.

Read your favorite textbook article at page 25, and then begin your program.

Let's examine the procedure in detail and what parts are needed.

Proceedings

To create a robot that uses one button to move, we need to create an loop to drive it.

To do this, first we need to separate the robot into the body and programming.

We will only set the robot to drive once you have pushed the button. To set the robot up to drive, press the button only once.

This requires us to first know when to push our button. We need to do this by entering our position in the computer.

It is possible to use 1+8+3=16 in order to determine if our device will move towards the second digit if we press the buttons on that digit.

Now we need to determine where the first Digit is. We do this by moving down and upwards in the program's view.

You can continue sliding down a number of columns until you reach the spot where your button must be placed.

The button is to be positioned and placed behind the robot. This can be done by sliding the box back in the same position as when we insert it and then into the position we need to position the button.

After placing the button behind robot bodies, the procedure can be repeated by pressing the button twice more. Once this is completed, we can return back to the source code to ensure that our robot is correctly programmed and ready for action.

What do robotics engineers do for a living?

They will take on human operators when the task becomes too dangerous for them.

What is the cost of running such a bot? How much would you have to pay an operator to work for such a robot? How much does a robot cost?

IEEE Spectrum took a look at the cost for robotic equipment, and this week we turned our attention towards the cost of the robots. This is the subject a new series called "Robot Economics".

We will examine robots that are capable of moving to offshore drilling platforms, heavy manufacturing machine, delivery robots, or even a R2-D2-sized toy.

Robots are an extremely useful tool, but there are still costs. This helps everyone, regardless if you are a manufacturer or a consumer.

The Robotic Petting Zoo is our first installment. It's a robot library located in New York City.

Dr. Dennis Hong of Toyota's Global Mobility and Mobility Engineering Laboratory gives a lecture about social integration for self-employed car owners.

"It's all about shared autonomy," says he, "while there's never a car at the center of this journey." Other people can also be involved. There is a bus, there's a taxi, there's a carpool, then you have a shuttle and finally, the vehicle.

Next issue will reveal how the robotics cost to perform certain tasks is lower than other tasks. Next time we will examine why fully-autonomous (robotic), vehicles are still an insurmountable distance.

We will be looking at how robots will affect society. That includes whether workers will lose their jobs or get new jobs. It will also show how it will affect cities and towns. And how adaptable society will be.

What is soft automation?

These machines were created naturally.

In the same way, robots can't compete with natural systems and their efficiency and precision in an ever-changing technological world.

Soft robotics can be described as a method in which a small part of material creates the appearance of a natural robot. Because they are made from engineered material, certain concepts of soft-robotics can be likened to biomimicry.

How are soft robotics manufactured?

These machines can be made using two methods. The first involves altering natural structures (eagle feathers or fish scales), and the second is creating or altering them. It is also possible to duplicate the properties of these natural structures using artificial materials.

India does not have any soft plastics that can be used to create soft robots. NASA and Defense Advanced Research Projects Agency are working to develop artificial muscles, wings and exoskeletons for global exploration. However, this is far from the truth.

How different are soft and hard robots from other robotics?

Contrary to a robot with an enclosed skeleton, a soft robotic has an open structure. It can be folded, exposed to, or contains an active fluid. This allows it to move and respond more freely to the environment.

So, if a mechanical force are applied to a normal robotic device, it either limps, resists, or deforms. While the soft robot is flexible and expands in one direction, it resists and resists. Soft robotics is not without its limitations.

Soft robots can be very useful.

These robots work in the same way as robots to pick up and deliver items. They can also be used as a means to transport heavy items to places that are impossible for humans. The goal is to be able to make robot arms and torso-striped arms using a 3D printing or sew-away technology.

ALSO READ: Indian defense industry is ready for the 5G revolution.

How does India's revolution take shape?

The largest company emerged from this revolution is actually one of India's oldest Kratos Defense Systems. This was in 2005. Kratos has made 4 Earth Explorer III vehicles specifically for landings onto the Moon.

In the coming months, there'll be more major startups - KEF Infraventures B.V. Company (a Bengalur startup) and Sindri Space (a Tata Institution of Basic Research-funded business that has developed a prototype for a 3D-printer).

Hemanth Kumar of Sindri Space spoke out in a recent interview. He said that his only size competitor was in. Sindri Space had a preseries A seed round in the amount of Rs 20crore (3.4 million), and its armor cost was more than one-year.

Rayform, an alternative startup, is working with Airbus Defense and Space to develop an inflatable module suitable for the Ariane 6 launch vehicle. This module is expected to have equipment far beyond that of the Ariane 6 exterior structure.

The company released a statement saying that they have "achieved success with many commercial assignments with material resistance exceeding 60t." This module will be produced in Europe prior to the launch of 2020.

What's the next step for the sector of space tech?

VentureBeat's vertical leader for defense and aerospace technology, Kishore Jin, said in an email that he expected the sector to grow rapidly.

He stated, "Currently the investments in the aerospace sector amount to approximately 30,000 ($4.2Billion), while space totals $20,000 (3.4Billion)." "In the next few year, the aerospace industry will grow by $120,000 ($17Billion), crore and $180,000 ($25Billion).

"Both sector are developing for one purpose - building test and investigating outerspace," he said. "We anticipate an exponential growth in the coming years."

Chapter 4: Programming robotics

Artificial intelligence can now teach robots physical tasks by using voice commands and motion sensors. In the video above, we see a Japanese warehouse's computer that learns how to load, retrieve, and return the boxes to people in dire need.

All this can be done with a simple webcam, and an AI algorithm. A webcam PC can communicate with it to give instructions for the robot warehousebot. Robot that recognizes people or objects. It basically has two cameras at its front and back.

The Bot's eye is Microsoft Kinect's. Bot's Bot can see the Kinect camera. When it is needed, it can face it to get the Kinect's information. It's not self-aware. It uses an algorithm similar to "Anavra", that tracks what he's viewing and uses that information.

The robot works well. It can recognize objects by the camera and so the robot begins with objects it already knows. The robot learns slowly about the objects and, if it recognizes one of them, asks Microsoft Kinect to display it.

The robot is able to pick up and place an object in a box but it cannot give it back. Because the robot recognizes both people and objects the boxes are placed in "service areas". The robot is then picked up by workers who take boxes from it and pass them on to the person to access their storage location.

If someone unloads containers, then the AI can't do its job. The AI struggles when boxes move. The robot hits its brakes often and is slow to move the cart. This makes the entire system slow to function. It is believed that only two people have succeeded in training their AI for this job to date: a human being as well as a computer.

These systems are intended to be able to "teach" AI systems each others. The AI program will do its own work and tell the bot what to make of the data. It then gives the instruction. The bots have to be able and willing to adapt to each others.

This research is followed up by AI. AI can then be used to learn complex tasks and eventually to make the bots work with others. Although it is beneficial for the researchers, this is not necessarily good news for us. RIKEN works on a variety of projects related to this topic, including cloud humansoids, telepresence machines, etc.

Is it possible to make a robots engineer?

Google has created a one-year-long program to help undergraduates learn how to build robots.

Google launched the code summer programme this summer to allow students to learn how to code robotics in a convenient setting. Their new knowledge can be applied in a paid internship within one of Google's five robotic groups, which range from a single project to a robot viewing system.

Micah Curtis of Tippie Business College at the University of Iowa describes himself as a robotics engineer.

Today's engineering students concentrate on product design. This is changing as the robotics and engineering industries become more important in both production and young engineering.

Students will work 10 hours a weeks in a robotics club, most often in the summer for three month. A dozen students work together in groups to create and program software. Then they visualize their robots by using a 3D computer-generated model.

Google has the highest salary of any summer traineeship, which can range from $6,500 to $7,000. David Hyun from Google is a software engineer. He recently said that this isn't the main purpose of the program.

Curtis stated, having last year worked for Google's robotics group, "We are trying making our engineers feel independent."

Google's goal is to help you build it.

Curtis stated that they were trying to build a team of engineers capable of finding these jobs and understanding what robotics jobs are. "We know that it's a multibillion pound industry over the long term."

What is robotics engineering and how can it be applied to your work?

Companies can benefit from the innovative solutions of robotic engineers who have strong knowledge and skills.

Ernst & Young's 2016 study showed that 74% planned to increase robotics-related spending in the next 2 year, while 73% of respondents expected increased revenues.

Companies often hire robotically-engineering specialists to develop self drivable automobiles, drones and automated production methods.

Who are robots and why do they exist? Who are these robots?

Czechoslovak scientist KarelCapek wrote in 1920 a play about "robots". The character of the robot complained that the human race had forgotten it was robots.

The 1984 novel by George Orwell was about a mechanized, automated society in which humans were replaced by machines and weapons-using robots.

But robots and other artificial intelligence make up an important part in the global economy.

According to IFR (International Federation of Robotics), there are roughly 3 million robots worldwide in 2016.

One of the primary purposes of IFR is to create a classification scheme that clearly defines robots, and distinguishes them form ordinary machines and other equipment.

A group of Oxford researchers developed a robot that could pick up objects just like a human being, and it did so without any cables in 2016.

Kela is the name for this model. It uses an ingenious technique using cables and electrodes to direct electric currents at specific parts of your body.

What are the main applications of artificial intelligence robots in everyday life?

Artificial intelligence is an integral part of making robots better. Robotics is the application of computer systems to aid humans in performing specific tasks.

AI has also been used in recent decades to help robots feel more like humans: think virtual assistants. On-line learning systems can be thought of as virtual assistants.

Deep learning was used recently to improve robots' ability to learn new jobs quickly and to speed up their completion.

What are AR's current robot applications for?

Augmented reality (AR), or systems that overlay information onto an atmosphere, is a way to make a realistic and more efficient experience. AR is a useful technology that is often overlooked in robotics because it provides a sense of 'virtual being'.

This can be used for many purposes, including:

AR can aid users in making information and objects easier to comprehend, especially when combined together with machine learning.

Enhanced Reality can provide navigation information. This can make it possible for the user to efficiently and easily navigate through complex systems.

You can use increased reality to give alternative information like instructions. This is much more intuitive than reading text on road signs.

AR can aid users in communicating information such as how an object works, how to explain complex pieces of a machine, or how best to install a new part.

AR is needed by some robots in order to learn and perceive their environment. Baxter for example requires enhanced reality to see their environment.

AR can also aid robots in understanding what information is already known about an object or situation. It makes learning new tasks much easier.

AR has been around long enough and it isn't limited to robots.

Machine learning and AR will be the main areas of robotics 2018. We'll be looking at more detailed commercial and technical trends over these next few weeks.

What is robotic Therapy?

Robotic Physical Therapy is a method of rehabilitation for those who have been injured in an accident or trauma.

Traumatic Brain Injury (TBI).

Traumatic nerve injury (TNI)

Junction trauma

Soft tissue damage (muscles, ligaments etc.).

Mobility, flexibility. Balance and control problems

Long-term physical strength increases

Moving/Motion in Repetition

Problems with postural instability and movement

Injuries can be caused by many different causes.

Trauma from an Accident

Trauma of falling

Modification resulting from physical activity

Frozen weather

Heart issues

Chronic disease

How robotics helps virtual reality (VR).

Virtual reality refers to the immersion into an actual, 3D environment. It can be experienced using a headset. A physical therapist can use virtual realities to create a new environment or recreate an existing one. Once that is done, he can experience the VR headset for himself.

She is a strong marketer and brand manager with strong experience in marketing. She has built and operated multi-million-dollar brands for the fitness and health industries and is the Global Marketing Director at Wi-Fit Corp.

Chapter 5: What is the future in robotics?

Most people wouldn't know what to do if robots entered a space.

A tree is a good place to look for a person if they appear.

If a robot were to leave a sandwich on the kitchen counter, most people would just stare at the blinker in disbelief and call the real human being.

These and many other basic forms of interaction between robots (and men) were discussed on Wednesday in a day-long panel discussion at The Rotman School of Management at University of Toronto.

The University's Institute of Creative Technologies was hosting the event as part of Milton Robotics Investment Conference. It wanted to highlight the great academic work taking place around the globe.

The Panel was hosted outside of the ICT research facility, which aims at both expanding artificial intelligence (AI), rapidly, and developing intuitive human robot interactions.

"You're talking not just about AI when this integration is possible, you're talking also about the future in robotics," stated Fred Turek from ICT executive directors. He was the one who had developed the IBM Watson computer.

Turek says that it is not necessarily a question whether we are creating humanrobot interactions.

"These technologies have reached a mature stage," he stated. "I think the problem will be over the next few generations how quickly these technologies mature."

Machines have been around for some time. However, ICT research suggests that we are only a matterof time away from making progress.

"Our goal right now is to build technologies that enable you to interact intuitively." This allows you to use your sense of touch and sight, sound, sound, and smell in order to interact with the technology.

Turek stated in the current stage that AI will combine with machine learning to create technologies that interact more intuitively and naturally with human beings.

The most interesting thing about robots is their ability to adapt to the speed of change in the coming years. He said that technology is changing so rapidly that it may be challenging to keep up within a few years.

"We are looking at some of these key questions: "How can we understand something in high-level ways and quickly control its interaction?'" "

The panelists suggested a number of ways to integrate this technology in their own lives.

Turek observed that technology would allow people to interact less with the things they need. A robot could be preferred to a person ordering or using a voice system to communicate with them.

"We have the technology to integrate these components. Then you can go in a bank or shop where there is no interaction with people."

Another possibility is to choose to interact directly with a computer rather than a person. If the interaction is short and easy, however, more tasks can be automated. This makes people more valuable.

Turek said that it is even more crucial when the machine is smarter, can process and analyze more accurately. "Having robots in your hospital rooms means that you can still talk with your doctor."

Where do robotics experts work?

According to a recent study from a leading staffing firm, robotics engineering is one of fastest growing areas in the technology industry.

Burning Glass, an industry analyst found that the tech sector has experienced a 60% growth in employment over the last five years. Two of most in-demand specialties are software and robotics engineers.

Marty Liimatta from Burning Glass, CEO, writes that robot-experts are in great demand and face wage pressure. There are many factors that lead to this demand.

Google and Facebook, Silicon Valley titans, have made substantial investments in robotics. Facebook is also promising to open a laboratory in Stanford University for artificial intelligence research. Microsoft revealed a new robotics project earlier this year.

Burning Glass says that robotics is a growing field with around 87,000 workers in the United States. This is almost 25 percent more than the previous five years.

Liimatta wrote that with rapid automation, workplaces as well as fast-food outlets are becoming more vulnerable to robotic substitution. "The labor markets for robotics are poised to explode in the next 10 year and will include a wide array of manufacturing, administrative, and commercial jobs."

Let's have a look at the results.

What does a robot technician do?

Robot engineers build robots.

David Evans (TechStand CEO) said that a robot engineer is designing robots to help people. He is now exploring how robots may be useful for people.

Evans is a robotics expert. A large part of Evans' job involves building robots for people. One example is a robot that can monitor and assist people with dementia.

Robot engineers may also develop robots specifically for certain applications. Robots that can carry items could be used in retail shops or airports.

Liimatta stated that the demand is strong for these engineers and will grow.

How can robotics engineers find a job in the field?

Evans stated there are two things that you need to do in order to get a job at robotic engineering. First, you can find companies that are looking for employees and they advertise. There are also universities that offer engineering jobs.

Evans said that Evans believes there are many companies hiring, and that universities are also hiring. "The problem they have is they don't have enough qualified candidates."

Evans stated that Stanford and University of California Berkeley are the largest markets on both the university and private side.

Evans said, "It can become quite expensive." "Many people (especially men) need to learn an entirely new curriculum. This can be very daunting."

Evans recently received a doctorate. Since 2005, Evans has been teaching and researching in robotics.

He claimed that he spent roughly $100,000 for his PhD. But, he received only $45,000 wages, mostly from work outside the university.

Evans explained that a good robots engineer can expect a revenue of between $95,000 to $115,000 per year.

What do software engineers do?

Software engineers write code to assist people. Evans stated that they are often the "grunt" employees in the field.

Software engineers are specialists in one area. A robotic software engineer may be able to write robot code that assists manufacturers. Software engineers are also able to specialize in web development, customer care, and other areas.

Evans said, "Job Demand is High."

Evans estimates that robot software engineers can earn around 108,000 dollars an year on average, about 30 thousand more than the average software engineer.

Evans stated that "Even if you do not have a direct connection to robotics it's important to realize that it still has a significant impact." It may not make money for you but it shows that your knowledge is valuable and helps you build your network.

What do financial analysts do for a living?

Financial analysts use currencies, stocks, and bonds to analyze trends and predict the future of certain stocks.

Evans said, "The job of a financial analyst has not changed significantly over the years." "What has changed over the years is the amount of data that computers have access to."

Evans says that financial professionals also analyze news, media and other data.

Evans stated that Financial Analysts earn on average $130,000 annually and can expect earnings up to $150,000.

What do the mechanics do for us?

Evans stated that automobile engineers can also be responsible for the design and development of new materials, manufacturing processes and other innovations.

Also, there are engineers in the machine shop that create the machines for the cars.

Evans says mechanics can earn around $53,000 per annum and can expect to make up to $63,000.

What do IT technicians do?

Evans explained that engineers who are skilled in computers can also work within the technological field.

They work on projects like websites and mobile apps as well as big data and how the internet delivers things.

Evans stated that IT workers can expect to earn $71,000/year and as much as $77,000.

What are people doing using fixed phones?

Evans said they work in mobile software.

Evans stated that they "build apps and collaborate with developers to create apps."

Evans said that repairs are on the rise. Repairs to mobile phones can bring in an average salary of around $22,000 per year, with potential earnings as high as $25,000.

What do physical therapists do?

The physical therapist helps people to feel better after they are injured or have had operations.

Physical therapists are also able to work alongside rehabilitation specialists as well as nurses, doctors, social workers, and even social workers.

"They're not the therapists we know about, but they're therapists working with people." Evans said.

Evans claims that physical therapists are able to earn $69,000 an year on average. They can also expect to earn up $88,000.

What are musicians doing in the music business?

Musicians are individuals who make and use instruments within the industry.

Musicians can be composers, writers, or performers. They can play blues or rock, country, pop, jazz, or blues.

Evans stated that a musician, who scores and collaborates with other artists, will be able generate $56,000 annually.

What is it that marine biologists do for a living?

Evans stated that marine biologists can help to develop solutions for ocean problems and ways to conserve marine life.

Evans said that Evans' work can be "really cool". "There's a lot of clever companies like SeaTac, who can help with anything from salmon projects to Alaskan projects.

Evans says that marine biologists are able to earn an average salary of $79,000 per year, and they can expect to make as much at $96,000.

What are the responsibilities of doctors?

Like mechanics and financial advisors, medical practitioners work in brain. They work in an operating room or with anatomy.

Evans stated that they are capable of doing a lot of great work but can't do it all the time due to their limitations.

A doctor can expect up to $200,000.

What does it take to be a Sales Representative?

Representatives deal with business to-business companies.

It allows companies to market their products and services.

People working in this field often work long hours and are constantly on their feet.

Evans claims that an average pharmaceutical sales representative can make $75,000 a year. They can also expect to earn up $94,000.

What are caregivers exactly? What are caregivers exactly?

Evans stated that "they're people dealing avec the elderly."

She also suggested that they might be responsible in caring for disabled children and persons.

Evans stated that "they're paid just to look after people".

Work can be done at your home, in the hospital or in nursing homes.

Evans said that "they are skilled in many subjects." "They could be nurses."

Evans explained that the ideal caregiver could make around US$55,000 per annum, according to Evans.

What is "technologist medicine"?

Evans said that the people in this industry were trained to manage diagnostic machines, devices or other equipment that support the health care sector.

The technology is able to help you diagnose, treat and monitor a patient.

Evans stated they can be nurses, doctors, technologists or other healthcare professionals.

Evans said, "These people have an avg. salary of around 50,000 dollars per-year and are likely making about 62,000 dollars."

Chapter 6: How can you start a robot club?

Adelaide University robotics is coming together to create a new campus club through THE ROBOKIDS. How will a club evolve? Can Australia be saved?

Robotics is gaining popularity across Australia with teams leading national robotics contests and opening chapters to support students in the FIRST program.

Robotics has taken a backseat in many schools. But the new association aims to "allow Australian robotics the chance to regain impulse" as well as to enable students to take robotics up a notch.

Professor Lachlan S. Stirling of School of Computer Science and Computer Science is involved with ROBOTKILLS. He says robotics can be a great teaching tool and Australia was late to the party in establishing a national robotics education program.

Stirling says Australia has a chance of catching up with the rest and gaining a competitive edge through robotic training.

"The problem is in robotics as well as in the coding, and coding of robots. It's a matter of getting kids interested at an earlier age," he said.

She states, "We have the ability to get people interested in robotics earlier than others and I believe this is important."

Stirling believes a club could be a good way of involving students in the "truly complex and important field" in light of the increasing use of robots in news.

He stated, "We need be the leaders in research-and-development in robotics because there so many aspects and uses we can learn,"

"This project is all about engaging students with robotics, inspiring them, and engaging them with an exciting field."

Stirling and Professor Christa Dahly developed RoboKIDS, an engineering program. RoboKIDS Club was founded on ROBOTKILLS. It was held at Adelaide University for the past four year.

This two-year program was run at Adelaide University. It consisted of various robotics challenges culminating with the final robot competition.

Young students were very interested in the robotics competitions. The six teams were self-employed and included service robots as well as robots for performing arts and Lego robots.

The RoboGames competition took place at the University of Adelaide last Saturday. It awarded $500 to the winner team, and a LEGO robot.

ROBOTKILLS's end does not mean the end to the program.

Students will have the opportunity to decide how they want to apply the knowledge from the programme to create an aerospace, industrial, and citizen robot - the "ultimate bot capable of solving human problems."

Stirling states that it is also about encouraging young people "to go to higher educational and to work for the robotics industry once their degree has been obtained."

"It is an exciting, difficult, and vital field of science. It is why we need start robotics in Australia and to develop an engineering talent.

EUROBOTICA 2017 will occur in Melbourne and Sydney in October. Robocraft is the theme for this year's European Championships.

How has robotics transformed production technology

Robots in every facility What has been the impact of these changes? Are there high-value automation processes?

Ian Michael heads Thyssenkrupp Production Solutions Ian is back from the US Annual Operations and Technology Conference. Arie van Meerten, CMO at SoftBank Robotics, and Robert Mead, CEO, both SCIO (Strategic Collaborative Innovation and Technology).

How do I get started in robotics

A few years ago I moved from California to Pennsylvania. Before I was transferred, I had no prior knowledge in robotics. I didn't have any robotics training at my college and I wasn't aware of robotics in the region where I worked. Prior to moving, all I knew about robotics was through TV and press.

Many people move into production to become factory workers. I chose another route. I was seeking something else, and robotic technology seemed appealing to me.

Before that I worked in a dental office. 3D printing was something I came across through that experience. I found it fascinating and I am a huge fan of the endless possibilities it offers. I discovered that it is possible to build your own equipment for a 3D printer.

How do these concepts are taught?

I purchased a 3D-printer for the first time, just like many others. After that, I wanted a robot. YouTube was my first source of information. The first book I bought was Robotic3D printing (University of Washington, 2005). I found it on Amazon for $37.

When I started to research the process, I discovered that most of the material in my book was available online. I started by looking for a site that explained how it works. There's a lot of information out there about 3D Printers. I was having difficulty deciding which 3D printing machine I should purchase. I didn't know which style I needed.

It is easier for you to move and make use your own parts. It is much easier to begin using a videogame controller with a controller specifically designed for gaming. It is also simpler to start using an industrial controller, such as a PLC, with a controller already designed to work in conjunction with videogames.

For example, if a piece moves with some movement, then it is not possible to just make a robot from that part. A robot like this one has gears or wheels. You will need to find a motor that matches the motion of the robot, or make wheels or gears that match the function of the control system.

How did it come that you decided to return home to Minnesota to start a business?

I saw a market demand. People don't buy robots only in their living room. They began looking for robots while working. I decided that I would start this business. That experience was already mine.

You state that it is important for engineers to have a solid foundation.

It is a great way for you to read a lot. You can find a lot of information online. The amount of time you spend learning and the speed at which you learn it will affect how much you can learn.

Over the past few months, I have learned many things. I am always learning new things. I have learned a lot about patent building, making companies lawful, and complying with federal and State laws. To be able to determine which laws apply to you business, such as when your robot stops operating at night or stops working, it is necessary to research these laws.

Happy Robotics - Would you like to work there?

It's been a lot of fun. I love making new products. It's also exciting to see the robots live. I love watching the robots live.

How to make a robotic arms?

Gerald Dunn (Georgia Institute of Technology) is a graduate student working with robotic hands to improve people's everyday lives.

This video has been optimized and optimized for mobile viewing through the BBC News application. You can download the BBC News application from both the iPhone App Store (and the Android Google Play Store).

What programming language do you use for robotics design?

TUG Robot is developed for some major consumer goods companies. The ROS API was chosen for our design platform as it makes it easier for engineers. However, we also have years and years of experience with every major robotics platform as well as systems for the industrial machine.

Zero, is the ROS-server's preferred language. Zero is the first ROS server to allow Zero to use any industrial language. However, we want to maintain the existing ecosystem for robotics languages programming. It is also compatible with all other robotics languages.

What is the biggest problem facing industrial robots?

The robots may not be everywhere. Robots may not be available everywhere at the beginning. The first reason is that robots still are not everywhere.

What is the distinction between production and sales?

Rowrobots are currently being produced, but they are still far from being able to do more complicated tasks such as speaking, making dynamic gestures, and complex dynamic playing. This is what we aim to achieve with the TUG robot house robots. Retail is completely another.

Manufacturing robots must be programmed to solve problems such as machine operation. If the task is too difficult, it will fail. But, automating repetitive tasks such labeling can be a great way to create a robot which can read the product and place it on a shelf.

Roomba is the ideal robotic vacuum cleaner for retail. It is a simple robot vacuum cleaner with some useful functionality. If you have multiple product categories you would like to include on your shelf, make sure it goes from the back into the shelf. It will then be able to navigate the shopping corridors.

How can you bring more robots to the retail environment?

Our vision is for robots to be used in the retail industry more easily.

This article was published under headline "Retail bots."

What are the laws and principles of robotics technology?

Computers have changed from being simply assembly line models to mobile robotics, computer viruses and artificial heart. Robots have greatly improved over the past 50 years.

Future plans for military robotics?

It is possible. "Why waste billions of dollars on developing driverless vehicles when you can have driverless killing robots do the exact same job?" According to the military

Although any techno-bureaucrat can quickly affirm to "crackthe code", which will result victory, that's not what the Terminator team was looking for. They had a plan to dramatically increase death rates in order to win the war.

These concepts, which look a lot like terminator-like ones, aren't even possible to execute. Although battlebots resembling a terminator could be the future of warfare, the area has many variables such as an aggressive enemy, insufficient weapons for all human foes, and many other factors. It's one of the biggest faults that could lead to the demise of the entire project.

This is not to suggest that the M1A1's system is not worthwhile. However, some industry experts say it is a great tank. It is not clear if the US Army ever had to purchase a Tank with the Terminator Warbot characteristics in order fight the enemy. Even if it was cheap, the overall budget would not have been significantly affected.

What is a class in robotics?

Robotics encompasses a wide range of technologies and could include driverless cars (or drones), wearable computing, and aerial drones.

Like other technologies, the use of robotics by business and public depends on its field. Empathy is the key element in all of these fields.

Research is abundant on how perception works and how we can adapt to it. It is critical to teach children to look at the emotions behind actions.

This is especially important as society becomes more fractured.

These are 10 of the most popular toys in 2016 according to the Register.

How do I invest in robotics

Robotics is hot because major players are creating new products and tools to stay competitive. Most people do not know the consequences of robots on our daily lives. Jason Hiner (our resident robot specialist) gives you all the details to help you invest in this rapidly developing industry.

SEE: How the artificial intelligence was constructed in the robots report by TechRepublic

Mobile Enterprise Newsletter BYOD. Mobile security, IoT mobile security remote support and IT professional smartphones, tablets and apps.

See also

Chapter 7: What exactly is robotics technology and how does it work?

How can robotic technology assist the visually impaired and blind

Credit: Colorado University

One might consider the remote control that controls a television or an automobile's navigation system as the technology that has allowed millions to lose their sight. Robotic technology can be used to control television and automobiles, and it can also assist visually impaired persons.

Technology is necessary to aid the blind and visually impaired in their integration into the world of today's society. Combining computer vision with mechanical robotics, vision enhancement can be achieved.

According to the National Blind Federation of America, around 30,000,000 Americans suffer vision loss. Of these, 17 million are completely blind. Nearly 2.1million are blind and 1.8million have total blindness. However, technology has begun to include assistive technologies like Braille screens, symbol language translators and closed captions.

Christopher Van Heusen of the University of Colorado Boulder's Department of Mechanical Engineering stated that robotics technology has been developed to enable blind, visually impaired, and disabled people to move about and participate in everyday life. This technology can also be used to correct vision problems in those who are not dependent on others.

Equipment for the mechanical trade

A robot is a device that can assist the blind and visually impaired. Robotically enabled assistive technologies can help people with daily tasks, such reading and driving. Additionally, they may allow them to offer services worldwide that will help them be more active.

Matthew W. Thomas from Rochester University has created the MagiVis. This robot uses a camera that scans real-world images to approximate the object. The MagiVis provides visual aids for the visually impaired and their families.

Another robotic device called the ReadingEase connects to a computer which can read plain text as well and braille. The computer can keep the braille section with most points at the top, which is what blind people and the visually impaired need. ReadingEase developed by University of Minneso.

Vision of the Computer

University of Minnesota currently is developing RightEye ObEN, which can assist the blind with basic insights. The robot is capable of reading both complex images and highly contrasting shapes.

Researchers plan to use computer visualisation to improve braille systems in the future for people who are blind, so they can read books.

What is robotics' first law?

University of Warwick research on human/robot interaction. RoboKind was his 2011 book. He explores the notion of a fourth group of robots that consists of both autonomous (cyborg-like) and human/robot interactions.

What ethical and judicial questions can be raised about human-robot interaction?

Researchers and academics have demonstrated various types of robotics, including:

Robotics for humanoid care to improve speed and accuracy of neurologic diagnoses.

Simulators based in agents designed to model a wide variety of human conducts.

Computer-aided teachers can be used to aid learning in math and other subjects.

Support robots for learners who are developmentalally disabled.

Basic AIs were used to create intelligent software to allow interaction between human and robots.

Social robots may be used for social assistance, such as older people and those with disabilities.

Robots with meaningful interaction will be able to support the elderly by being interested in their humans.

Ethical problems arise when robots are not used in a predictable and predictable manner. Emergency, police, military, and care and educational problems may arise.

Recent technological advancements in robotics raise concerns about robots' potential replacement of human caregivers in the home and in nursing homes. Recent studies have revealed that occupational robots are being used in more than 40 occupations.

These ethical concerns are raised in relation to robots' roles in war. Autonomous robots raise significant questions about the imposition of lethal force.

Some of these ethical concerns have been addressed. There are increasing discussions on the potential applications and limitations of certain topics.

Education – Professional ethics regarding interaction between student and teacher

Research and design – Research ethics in the context of creation, reproduction or learning ethics

Commercial ethics -- ethical issues relating to the design and use robots.

Health and Social Care are a growing list of ethical concerns and potential uses and restrictions

Social Studies-Design, ethical assessment, and education of robots for community and social studies

All of this and more.

What does International Robotics Federation mean?

The International Robotics Federation announced the first robot ethics statement.

The declaration states that

IFR won't tolerate any measures that would jeopardize or menace the safety or wellbeing of a robotics operator.

A robot user should assess the potential risk and take appropriate precautions to avoid these situations.

Remotely controlling a robot from outside can be dangerous. It could also be considered an attack or battery.

Risks and accidents could result from robots used in cleaning industrial jobs.

Robots that do not work properly can be removed. This could cause additional damage.

Every decision that you make regarding the use and maintenance of a robotic device should be taken with a clear, informed mind.

So, what do you do?

As previously mentioned, organizations provide guidance and information regarding the ethical aspects and social ramifications of robotics.

Ethical Robotics Society (Ethical Robotics Society): Works on the Ethical Robotics.org site and hosts an annually international conference on robotics ethics.

What is robotics' second law?

Robotics second law: Have you ever thought of it? See our previous article robotics first.

"A robot can't injure someone or allow them to be injured by inaction," is the first law governing robotics.

This law was first mentioned in Isaac Asimov's 1951 book Robots and Artificial Intelligence.

Rob Gonsalves is a robotics scientist at Carnegie Mellon University. His 2011 article Science Advances, which was published in Science Advances, shows that it's possible for a robot injure human beings.

Let's say it's 7 December 2016, one year before Donald Trump took office.

Helios, a MIT-research robot, was the robot. Helios was asked in a game simulation to take a glass water and place it on the dining table. It could pick it up, place it on a table, or use its fingers.

The robot could take the glass out of the container when the experiment started but couldn't place it down until its arms tired.

Helios eventually learned to pick up the glass, but it was an arduous task. He would set it down for each glass and then run for just a little. It went on until the water ran out of the glass.

If AI's future is like that of the robots at the laboratory, then Helios could be considered very good.

What about future robots?

If you picture a robot walking in real life and performing tasks. You can also imagine him being able quickly to figure out a task.

This is robotics' second law.

Future robots won't just learn how to pour a glass of water (see the previous article), but also how to play soccer quickly.

Robots can complete many tasks at once so they don't necessarily need to know how to do every task. However, they are capable of doing a lot at once.

But how can you be certain that robots will soon learn to play soccer?

This is what I learned in a speech delivered by James Hendler from MIT. He explained that the 2nd robots law is best for robots that are already in use.

This means AI is already integrated in many industrial robots. AI can even learn how football is played.

We do not know how the next generation or similar games will learn to be football players.

Hendler suggested to robots that they learn about the rules of each game.

My belief is the scope of play can be extended to all dimensions, including space, virtual, operational and physical.

A second dimension could be climate. This dimension could see rain. Maybe a robot will even learn how rain works.

In the future, robots might be able teach themselves football. Perhaps they could play football and learn chess simultaneously.

An algorithm, for example can be used as a way to learn football. It then can be used to learn how you play football.

If we can increase the playing field (for instance by adding a third dimension of skills), that is certainly one of the benefits of evolution.

The context may be the last dimension added to the playing field. Robots' ability to move, inspect, and possibly even destroy objects will depend on knowing which objects can be considered harmless and which are hazardous.

They can also learn these things from their play with objects and games.

All these elements increase the complexity of this game and make it more challenging, quicker and more enjoyable for all.

What is robotics' third law?

The Terminator, 1998's film, is a good example of what we are talking. If not, you can find it in the 'third law on robotics'. It says that robots cannot harm or permit humans to harm through inaction.

The law states: "A robot must obey human orders except where such orders conflict with the first."

The first robotics law' states:

"A robot must adhere to human orders, except in cases where the orders are inconsistent with either the first or third laws."

The 'Second Robot Law is:

"The robot must preserve its own existence, provided such protection is not in contravention of the first and second laws."

The robotic 'Terminators,' which will be created when humanity fails, will continue to grow. As it was predicted, humanity is going to eventually fail to save its own self - leading to a new robotic Holocaust.

The franchise "Terminator", which is almost directly inspired by the Dune' series, shows that the third law' of robotics plays an important part.

Frank Herbert (the science-fiction author) has stated that "Dune," specifically the Nastarian race, is his personal interpretation of religious myology.

The Nastaran religion is one of the most powerful religious groups in 'Dune. It is characterized as a feeling of unanimity with all things," said he. "In fact, their entire community lives in a tight circle in a symbiotic relationship with a plant.

Nomenklatua, also known as "the Wormflower", is a poisonous that 'neutralizes' all toxins in biological and machine air.

The Chinese tribe's religion inspired "Dune", the fictional religion. Many Chinese tribal beliefs were actually inspired by the original Christian lessons.

One of the Chinese tribal religions, also known as the diao yi, is often thought to be the "skinshedding"

"Some faiths that date back to the 7th Century are capable of removing skin. Therefore, a person wearing scorpion skins will be seen as hostile and more likely to hurt them. On the one hand, a person with pale skin is weaker and less likely than others to harm.

Herbert said, "There are certain organs that are inextricably tied to the nature or character of your character." "In simple words, a person becomes another person when he sheds his skin. A Christian can suddenly become a Mormon, or go from a straight and narrow Christian to a debauchery lifestyle.

It doesn't matter if these films are Francis Ford Coppola's "Dune" or "Terminator," it seems like they have at least one common topic: technology and its impact on humanity. Those who embrace it can have horrendous consequences.

It is a subject so complex that it could have its own Third Robot Law.

What is robotics meant for children?

At the Sonoma Valley Library, on Saturday 17th March at 11 am, the Valley of the Moon Audubon Society is going to present a special program on the future of children robotics.

"Robots for Children", part of "Backyard Worlds - Planet 9". Planet Nine was discovered in 2016 using technology for school children.

The lecture will be given by a guest instructor as a fantastic opportunity for kids learn the basics of science and technology engineering and math (STEM).

The Planet Nine exhibition, currently free at Sonoma Valley Library, is open to all until Sunday 22 April.

Highlights from Sweet Spot's Jazz Singers Program

The Sonoma Valley School Jazz Orchestra performed hard bop, swing and more. Delicious hors d'oeuvres were prepared by Ed Keenan. Wine pairings were provided by Mauricio SOM.

Som, Manzanita Rock's owner, organized a private dining experience for the band members along with Cheryl Som as Cheryl's food & wine manager.

Chapter 8: Where are robotics used in the real world?

As robotics technology develops, there are more industrial experiments that allow people to work alongside robots. Robots have taken on tasks previously impossible and humans have been able to do more complicated and complex jobs that were previously very difficult.

The number and quality of jobs in the US declined dramatically in 2018, as robots became more advanced. We can all play a greater role in our workplaces, because robots are capable of performing more complex tasks faster and more efficiently.

Robotics investments have led to an increase number of automation companies operating in all of their offices. The impact of this technology is incredibly significant: how we work, what tasks we complete and how many job openings are available, are all drastically altered by its development.

The rise in wages has been caused by robots gradually taking on more work over the past few years. Companies have been able go forward and become more automated. They have also been forced by law to pay higher wages to employees who have the ability to adapt to new working environments.

Many businesses now consider the importance of maintaining employees and paying them higher salaries. With the automation of certain jobs, one of the biggest challenges for companies is that many employees want to feel safe and secure.

This change could be attributed to the lack of skilled workers. The majority of jobs, such production, are more difficult and require more analytical skills and analytic abilities. Companies in manufacturing are developing new methods to stop workers leaving due to the decline of jobs.

Many industries have been negatively affected by robotics. Businesses find innovative ways of retaining skilled workers. Kiva Systems makes use of technology to do warehouse work. This type of automation can allow for computerization.

Although qualified employees are vitally important, it is natural to see a decline of human employees in the workplace. The value of specialist and specialized employees in their businesses will increase and they will be able more safely and effectively alongside robots.

How can you get a job at robotics?

New data indicates that people are looking increasingly for robotics (AI) and artificial intelligence, which is at the very least the hope of robotic managers hoping that the boring area may avoid the dark side Internet, especially porn.

Ava Robotics CEO Tom Ingersoll explained that robots are joining the sex market because there are more opportunities to them than security threats. Problem is that robots will not be available for women.

Ingersoll said, "I don't believe that there will ever be a bunch'sexrobots' for sexual pedophiles. However, if it is clear that my housekeeper was tired and had too much to drink, then I can hire somebody that's a bit more friendly for me."

The robots do not have to be gentle. Robots are made by men to fulfill fantasies and perform certain behaviors.

Ingersoll considers the sex industry to be an area where the development of robots makes sense.

He envisions female robotics that can adjust their facial expressions to meet different desires. And male robots for men that can change their breasts and color their hair.

Ingersoll envisions Ava as a web personality. Ava is the first product that his company has produced. He envisions Ava being a character that can be created by her friends and interacted with so that they can communicate with their true friends.

He sees it in a way that robots can help each other learn new tools and applications. He seeks to make robots less cruel and more adaptable to new and varied situations.

He is aware that not all people will agree with his positive views of the products.

Ingersoll explained, "We are facing a seed that is out here, not only in the internet but also within the robotics industry."

Ingersoll says he hopes more universities will offer more courses in robotics. Nowadays, artificial intelligence (and robotics) are two of the most-popular subjects In computer sciences at universities. Ingersoll failed to realize the absurdity of his idea when he tried selling it to many universities.

"They were like, "You are joking?" They wanted me to teach robotics while the porn business teaches people how sexbots are made.

Ingersoll admitted that it is unlikely that sexbots will ever become a reality, but that they are likely to increase in the face of security threats.

Ingersoll said that "I don't see them getting there because I don't believe people's going to want to sleep on machines."

Ingersoll indicated that sexrobots may not prove safe to be used by one person alone. They may lack impetus control. Or they may cease reacting to verbal commands. He suggested that sex robots might be unable to differentiate between a deep touch from a "strange or crazy one"

He explained that sex robots can be more like people and become more interesting and useful.

Ingersoll explained that sex machines, like any other technology, will most likely be embraced by the public.

Are you looking to play a futuristic video game?

Robotics do not have to be limited to the military. They can be used in science, industry, and manufacturing.

What is a Robotics Team?

A robotics staff is made up of students with an interest in engineering. The group works together to create fun projects and learn more about the field. A team can be formed at your school or club. No previous robotics knowledge is required to form your own team.

What is robotics and how does it work? What is robotics?

Roboworld consists of a 3D design competition. Each team is assigned a subject, and often has to collect pieces to finish their design. Each team member designs and tests its robot and compares it to other teams to determine who can design the best robot.

A few weeks prior to the contest, teams will need to select a topic and design a robotic device. Each team must also make a video explaining how they plan to win the Roboworld Challenge. Before the team can compete, they will need to have a battlefield and an extra hour to test their robot.

What are some of the benefits of robotics

Learn how robotics may improve your experience.

Robots will become more integrated with humans and work in tandem to improve employees' experience and knowledge. These robots can combine artificial intelligence and sensors with speech control technology to improve productivity, reduce errors, and free up time for higher-value activities. These robots have become more affordable and reliable. Companies can now invest to improve productivity and customer service.

The RBC Mobile Robotics Forum will keep you up-to-date on all developments in Robotics.

The Royal Bank of Canada is an international financial institution that delivers exceptional performance through a principle-driven and goal-driven approach. We are proud to have over 84,000 employees that share our vision, value, and strategy. This helps our customers to thrive and our communities to prosper.

We are proud support a wide range of community initiatives through grants and community investments.

What are some of the disadvantages to working with robotics?

Boris Groysberg was President and CEO at Robosoft Consulting Ltd.

Recent research suggests that robotic labour in Australia is cheaper than human labour. In all areas, productivity exceeds cost savings of PS2.3 billion. While the market is experiencing a slowdown in robotics, technology in general is rapidly gaining in prominence.

It is increasingly in demand

Robots have become more popular in Australia. In the automotive industry, robot sales have increased by 35% in 2016, and they are expected rise by 42% in 2017.

What's the secret to this surge?

It is more than technology. Therefore, the supply chain has to change. We have introduced flexible manufacturing technologies and increased 3D printing. For this reason, many firms have begun to outsource work to low-cost countries in order to be able to compete on the global market.

The automation process has its costs, but these are offset by the lower price of the robot, especially compared to the human worker.

Australia's largest manufacturers have begun investing in automation to remain competitive. Australia is no more the fifth most important country for robots.

However, there is still plenty to do.

Australia faces many challenges regarding manufacturing robots. As of now, the majority industrial robots are cheap, so it's more cost-effective to hire than hire.

The machines are often sold on an unprofitable basis, either because the customer cannot afford them or they have a lot of machines. Therefore, we pay for machines not used by the manufacturers. In most cases, this increases the cost of operation, such as wages, electricity, and is usually higher.

There are many unanswered issues, such as what happens after technology is sold. What happens to these robots after they are sold? The fact that robots still make little sense in Australia is due to the low number of them.

Develop and improve education and training programs

Companies must adapt to new technology and learn how to use robotics. Most manufacturers will not be capable of using the machines they have currently, so each machine has a period to allow for adjustment.

But, they can be more flexible about using machinery and reap the benefits of training staff to understand how it works.

There are many great training programs for humans.

A company that wants to use the telepresence robot must, for example, understand its needs and the types of people with whom it can interact in order to ensure that it is effective in the environment.

High-skilled employees will be needed to train their current employees how to use robots.

Standards and implementation agreements

For guidance and support on standards, companies should look to their government.

There are many robotic manufacturers who live in countries other than Australia. Therefore, regional standards must be established. The number of robots in every factory is increasing. It is therefore important to ensure that standard operating procedures are available for each location so that the robots can be used efficiently and effectively.

Establish appropriate standards

Although Australia's current robotic manufacturing industries are reliant on cheap robot producers, this isn't the case in Japan and Europe. There, standards and implementation become more important.

Another challenge is the fact that robot makers often use different design standards, even though their products are all designed for the same end-user market - that's, manufacturers.

Companies should be involved in setting standards so that industry can achieve the same quality.

The question we need to ask is why are different standards for robotics necessary? In many cases standards only serve a small purpose.

The federal government should provide the financing necessary to include the industry in the development of standards.

Develop an upgrade scheme

It is important to be alert to the dangers of promoting new technologies.

New technologies are often introduced to market without any level of validation. Rapid deployment of new products can lead to the rapid deployment of other products with different interfaces and power supplies.

This means that you can get more support for your upgrade program, which can result in higher costs.

Many new technologies replace the entire function and functionality of a product or system. Hardware can become obsolete. The product may become obsolete in a matter of months or years.

It is essential that systems are supported by the right amount investment.

There are many Australian issues that need to addressed.

In order to provide enough support for your robot, and to reduce its operational complexity, you should plan ahead.

It is vital to cooperate with regional manufacturers to ensure they understand the correct robotic support process and procedures.

When the robot must be upgraded, it's important to validate and design the process so the robot producer can easily upgrade, understand what investment is required, and plan the change to maximize the cost.

Standardization should be considered as part of any upgrade plan.

Settlement

Robots will definitely have a significant impact upon the future of Australia's manufacturing industry. Industry and government must collaborate to ensure that robots are as efficient as they can be and that they are developed and maintained.

Who is the father in robotics and automation?

Today we live in an era where robots are available and offer, but these machines are more "balloons" than "living organisms." Although many scientists have tried to crack the code of robotics, none of them are closer to a functioning model than the classic HAL film of 2001: A Space Odyssey. A Space Odyssey.

Paul Hardaker, one of many robotics engineers today, believes that only a few science fiction films have ever come close to Paul Hardaker's imagination. Hardaker has created a scenario that robots will be able to organize a union strike in order to shut down Europe.

Hardaker said that robots aren't coming because they're helpful, but because the physical laws behind robotics have allowed us to change our current design.

Robots go beyond the production floor. Siemens, an international shipping company, has developed robots to climb escalators as well as unpack and transport food for their managers. This is not scifi fantasy.

But will robot technology ever really make its way into our lives? Are we really on the cusp a technological revolution, one that transforms our lives, TV and the internet, changing how we live and think?

Robot killers. Drones. Murderers.

The Future Of Humanity Institute, which is a group composed of robotic experts, promotes the notion of robot revolution. They describe themselves "maturing decisions-making, democracy, risk communication and computer literacy" as a way to fight against any machine uprising.

2011 was the year when the institute conducted an analysis of robotics' progress. The results showed that machines could "maintain critical jobs" within less than 50 years. It also suggested robots could in the future be used to fight terrorism. It was based upon the fact that human emotions and brain functions cannot be replicated with ease.

The Committee for Skeptical Investigation, however, found that the Institute's prediction was flawed. In January, CSI found that Mark Rylance a leading proponent in robotic dystopia did not review the work of top academics that countered the idea that robots could be close to reproducing the situation.

CSI performed its own robotics analysis and concluded that, in the short-term, the fundamentals behind robot function will not change. That people still need robots to complete their tasks, however. These machines are, in other words, still as intelligent, in their current state, as a dog-collar.

Practical issues

The Institute of Physics provides a more hopeful view of the future of robots. It believes that robots will have the ability to be useful in future nuclear disasters to help them locate and clear out nuclear waste.

Robot design presents many challenges, including "specific" materials and components. This is to ensure that the robot can function in extreme environments, such Arctic conditions or deserts, and that it can be repaired and rebuilt when it breaks.

It is also unlikely robots will make a significant impact on society as it stands today. Conservatives are already moving away significantly from low-skilled, labor-intensive manual labor for jobs that require greater intelligence, higher skills and more service.

A continuing decrease in employment growth will also occur, meaning that people will not have the option to retire or stop working.

Working and working together

One of the major issues in the CSI Report is the lack collaboration between scientists. Google spends a significant amount of money on robot development, so much research is being done in robotics. Unfortunately, researchers are not engaging the general population.

Groups like Future of Humanity Institute seek to devise policies to reduce the number of human injuries and accidents caused by robots.

There is evidence growing that robots will become more common in the future. This is why the CSI report has not convinced the public. A recent study revealed that 50% Americans doubt that robots can replace human beings in 10 years.

One of the concerns is the publication of a few "sceptical" materials on the Website by the group. One article is entitled "Why do human beings need attention to robot technology?"

The question is not about inventing robots. But, it's important to know when. We know that key events can influence human development. One of the major events over the past 200 years was the Industrial Revolution. It created new industries and changed the jobs of people.

It is possible to apply the same approach to the emergence of smart robots. These machines will be able help create jobs and alter the nature human work. We shouldn't be too pessimistic. But, we can prepare for a future in which robots are integrated into our daily life.

What is the average annual salary for a robotics engineer

It is not difficult to identify what the bottom hilt looks like. Low-skilled, low-wage engineers make a fraction compared to the top ten. The top tenth percent of the population are highly paid (i.e. Over twice the average salary of Silicon Valley software engineers, but the real cash outflow is in consulting.

It is interesting to note that 11 of the 13 people who fall within the consultancy category are consultants. While some may have other work in the industry while others are listed on this list, it is clear many of these are consultants. According to LinkedIn reports, Silicon Valley's average software engineer in 2016 will make $103,234 while consultants will make $163,000.

First, there is a significant wage gap. Second, high-skilled professionals have the highest salaries. They are also those with years of limited experience. Third, they will likely work in Silicon Valley, where engineers are available.

Five of the top ten consultants were interviewed for jobs in technology. It is also evident that consultants are tech-driven.

Surprisingly compared with the total employee count, the top 10% seem to be very low. There are less than 100 employees at this level.

These top companies have the lowest 50% wages, so there is not much room for them. The bar chart appears to show a flattening of these lower payments. However, there is still a gap between the average software developer and average consultant.

The wages for the first three positions are so high, it's not surprising that the average software engineering who doesn't work in a technology firm earns more than the average consultant.

You can see that the median engineers (who aren't advisers) earn less than those who work as consultants.

We don't know who the media engineer is but we know that Uber's Silicon Valley medium-level software engineer works there. Quora, the median consultancy engineer, talks less about startups so this may explain why there is such a large gap.

This picture shows the valley where the richest are the most prosperous, while the poor struggle daily to survive. The top is becoming wealthier and the rest are struggling to survive.

What is the answer to this problem?

It is important to push for social changes in order to make all workers more equal.

Research shows that benefits like healthcare can have an indirect impact upon wages. The government must still make it easier for people on the climb to success.

We cannot expect companies or other businesses to do this. We must start by forcing them to open up their supply chain more to external labour.

But it starts with us--the businesses whose employees work harder than anyone else to make their company successful.

What kind of education does a robot engineer need?

Robotics Engineers are engineers in robotics electronics and math. The way robots are constructed has a huge impact on how they learn and function.

Robotics is a rapidly-growing field and was voted the best job at work in America in 2018.

Todd Briand works with students from New South Wales every day for over 15 years. Todd was a program coordinator for the School of Engineering and Games and now he is the coordinator for the University of New South Wales' School of Computing.

Todd says that technical knowledge in education is the most important for robotic engineers. This means most programs incorporate both practical and theoretical learning.

Todd says this allows students to grasp the physical processes involved in robot design so they can understand math and feel intuitive regarding the actual system's processes.

Themes include robot and artificial intelligence as well as scientific and technological science, education, and human interests.

Chapter 9: Why robots are important

We focus on innovations in robotics, primarily in areas that have an impact on the lives of the disabled. We think mobility tasks are simple and so we are creating robots of next generation to do these tasks.

How My Robotics solves global problems with disability?

We believe these robots will soon be widespread in use, and can be integrated into daily life. This will help speed up technological progress for the disabled.

MyRobotics has created an open-source robot personal robotics ecosystem.

Many designers around the globe use the WeRobot Robotic Platform. WRP offers many options for accessing autonomous robotics technology. WRP also includes open-source robots that can easily be modified and tailored for specific purposes.

MyRobotics ML Library is a collection of components that allows for the identification, movement planning, and manipulation robot objects. We want to distribute these components to the public so they can make their own robots.

The Open Body Project also aims to make technology more accessible. It will create a class for robots that can detect their environment, interact, and carry out day-today activities.

MyRobotics has a young team that is open-minded, motivated and excited to work on global disabilities issues. Currently, they are looking into ways to improve WRP/ML.

MyRobotics aims to bring robots to life.

We believe that robotics has the potential to make a huge impact on the lives of millions with disabilities, if they work together.

Roko Robotics (the first branch of MyRobotics) is the company's first. It will focus on developing autonomous human/robot interactions for wheelchairs and mobility equipment.

The Group is also interested exploring MyRobotics' cases-based applications and robotic ecosystems in fields such as agriculture, education, and health.

Our staff

Six students are part of a group from IITRoorkee, an Indian engineering college.

Dhruvaj Suri Suri

Dhruvaj Suri is a core team member. Dhruvaj Suri has been involved in robotics research since 2013. He has published papers in various international academic journals.

Dhruvaj's interests include artificial Intelligence, interaction between people and robots, machine learning, computer vision, robotics, and machine education.

Adi Rai and Prof. Bharath helped him. All IIT-Roorkee student who are involved in robotics.

Dhruvay has a passion for learning and is a self learner programmer. Dhruvay's expertise in robotics, computer science and programming has enabled him the creation of MeeBot, which allows users teach and test virtual robotics.

Singh- Alexandra Vaad

Alexandra is a self learned software developer.

Machine education, robotics, as well as intelligent robots are her interests.

Alexandra's work in research on human-robot interaction was recognized by several national and foreign organizations.

Alexandra, who is a robotics student at IIT-Roorkee, is mentored daily by Prof. Shailendra.

Robo Bike, which teaches people how build their own robobikes, is also her co-founder.

Sharma Priyank

Priyank loves robotics and cognitive sciences, so he is now a MyRobotics program developer.

He has published numerous papers internationally in academic journals. He is also a proud member IMT Manesar India's leading robotic society.

Priyank is one Roko Robotics founders.

Singh Sanghamitra

Sanghamitra has published many papers in academic journals worldwide as a Ph.D. student from IIT-Roorkee.

She is currently a MyRobotics leader.

Pranav Mistry, all IITRoorkee Robotics students has inspired her.

Sanghamitra, a core Roko Robotics Membre, was honored.

Manjule's Shibin

Shibin is a MyRobotics engineer and is fascinated with robotics.

He has published several papers in academic journals around the world and was recently selected by the European Robotics Laboratory Vienna to participate in a summer research programme.

Shibin also is a core Roko Robotics participant.

Thakur Rinki

Rinki is currently a IIT-Roorkee engineering graduate.

She has published many papers internationally in academic journals and was chosen to compete at national robotics.

Rinki is also the co-founder and chief architect of Robo Bike.

Srivastava Rori

Rori, IITRoorkee PhD student is passionate about robotics.

She has published many papers at international academic journals.

Robo Bike's Rori has been a key member. She helps users build their own robobikes.

What is stem technology?

Modern medicine faces a significant challenge in recovering stem cells from patients. Since only one sample can produce thousands, it is difficult to determine how long and complicated the process will take.

Therefore, transplantation is a concept that physicians are not using enough. Currently, patients are treated mainly with autologous tissue therapy and autologous bones marrow transplantation.

Instead, stems cells may be taken from the patient and returned in the bloodstream for treatment of a variety of organ disorders such as stroke and heartattack, Parkinson's and spinal cord injury.

Stem cell treatment is about the ability to extract, transport, and replace stem cells within a patient. Stam cell treatment is the process of transferring stem cells from patients to their bodies. The cells can then grow and be expanded to treat other illnesses.

The enormous potential stem cells have to improve treatment options. However, it is not easy to obtain stem cells from patients. A joint team of scientists from Lawrence Livermore National Laboratory's Stem Cell Institute developed a method to remove blood stem cells, which allows for clinical treatment with stem cells.

The current requirements for therapeutic transplants require large amounts of samples. They also have low yields. Mark Chesley, lead researcher for the project, said that we must take the whole body out of the patients to create enough stem cells. We then isolate them from them, harvest them and re-inject them."

Chesley is a member of LLNL's high performance genomes resources at Stem Cell Institute, and a core fellow of LLNL's LLNL's reACT program.

LLNL researchers will collaborate with Alga/, a biotech start-up company for academic and clinical purposes, to optimize, extend, and develop this technique.

Chesley said that his platform was able to harvest stemcells from blood in 15 minutes and could prove that it can reliably collect stem cells from adult human skin.

Clinical success and business opportunity

Chesley said that they were able show how efficient and scalable this method is in multiple locations. This makes it highly practical for academic and clinical institutions looking to use stem cell-based therapies.

The team currently works on clinical studies, with ALS patients participating in clinical tests at the University of California Berkeley. The initial cohort of patients produced promising data. They showed significant improvements in their clinical and quality of living measures.

The equipment uses the same method to repair the backbones and bones of mice and rats, which shows that there is clinical potential to transplant blood stem cell to treat backbone injury.

Chesley stated, "Especially compared to other strategies for the clinical use like stem cell donation and autologous transplantation, the technology has potential to be an industry changer in regenerative medication."

Chesley and his staff also make tissue and organ-specific cell product on their platform. They use this platform to drug test, regenerative medicinal, toxicology, human tissue Engineering, and toxicology.

This work was funded by NIH (National Science Foundation), UC Berkeley and alga7.

National Laboratory for Lawrence Livermore

Lawrence Livermore National Laboratory(LLNL) is America's most prestigious national engineering and research lab. The U.S. Department of Energy LLNL serves national security through conducting research in a wide range of fields such as cancer, lightweight, national security, energy and security, nuclear systems and nuclear systems.

What is the forecast for the robotics engineers?

Police officers now have a wide range of tools to help them identify criminals and people who are missing.

While artificial intelligence technology has revolutionized society, most people aren't aware of its impact on their lives. Artificial intelligence can be found in many of our devices, including our smartphones, watches and even drones.

Some AI systems are superior than others. AIs are capable of driving cars via Google Search and Google Maps. But, the fastest benchmarking supercomputer in the world with 93 teraflops and 17.5% iPhone X-capability will only be made available to the AI System by 2019.

We are where we need to be in the global robotic revolution.

Most of the world isn't prepared for full automation. The jobs of thousands of bankers, assistants to banks and store clerks have already been lost. Some experts predict that robots or automation will take away up to 15% more jobs over the next 20 year.

Britain is widely thought to be one the most advanced and sophisticated economies on the planet. London has a flourishing robotics and finance sector, while Oxford develops the next generation AI system. It is essential to quickly make progress to avoid any major changes over the next decade, as we wrote in our paper, I Hate F&B Robots. Do you?"

The Philippines is one country where robots are growing at an incredible rate. The government is also investing heavily in the area. Although robots have been at the forefront in consumer electronics for a while, they are still used on the consumer markets in 2018 to make consumer products. Robots also play a crucial role in agriculture providing crops that can grow.

What is your job outlook as a robotics technician?

Myeisha Franklin talks about the future and robotics technicians

I had the privilege of speaking recently with Myeisha Franklin (industrial expert) about the future for robotics technicians.

We discussed challenges and opportunities in the field, what skills are needed today and how to improve our education to prepare for the workforce of tomorrow. Myeisha heads Optimus Robotics. They design and build robotics in the field.

Q: What is current status of robotic engineering industry in Canada?

A: Technavio's robotics career is growing. Technavio found that 17% thought they would experience the largest growth in the industry over the next five to five years. Robotics technicians can be considered the driving force for the future automation industry.

Industries around the globe have made significant advances in automation technology over 15 years. This field must continue to train and certify qualified personnel to support the advancement of robotics technicians.

Q: How is automation evolving? And what about the education system for computer science students?

A: Computer science and STEM have become more important. The future success of the robotic industry will depend on software maintenance and software development. It is also important that students learn how they can troubleshoot, and repair software systems. This is an industry that is currently in crisis.

While many schools offer programs that provide practical experience for students, most do not offer software maintenance. This is an area where students must be educated in order to keep current with changes in the industry.

Schools and companies need to ensure that their students have the necessary computer science skills for future employment. Students should not only learn code, but also learn logic.

It is crucial that students have a foundational knowledge of how to solve problems.

We need to be competitive in this industry by having well-trained and certified students, employees with the required skills to support automation system, and those who will be receiving programming and problem solving.

Q: What technological state is the curriculum in the present?

A: Many robotics training programs utilize very basic IT courses. They are often called "ITIL" because they teach students the "ITIL" or "Information Technology Infrastructure Library" curriculum. It focuses on programming, hardware development, and technical help.

These courses teach pupils the different aspects of robotics systems engineering, without giving them a detailed understanding. In the past robotics programs were introduced in classrooms using pre-packaged curricula.

Purdue University and other companies have been pushing for better interaction between human and computer computers. Covenant, West Lafayette, and West Lafayette are integrating computer scheduling into their curriculums.

Training courses are provided by companies to teach students core software building blocks. Teachers do not use these tools.

Although educators are not likely to use the software, educators should be aware of how it can be used in their classrooms.

What is Automation and Robotics?

Robotlogy and automation can be described as a term for various methods that people use in order to repeat tasks that require computer processing. Although this sentence is frequently overused, it is true in the main sense. The exact meaning of the sentence can vary because there are different processes for each technique.

When we speak about automating employees, the sentencing is often performed interchangeably. Because they are so distinct, this is not a good idea.

Automation, in its simplest definition, is the use technology to accomplish tasks that would have previously required human effort. Self-service checking is an example of this automation principle. It allows the customer to bypass long lines.

Robotics can be a very different process. Autonomous cars can be programmed to navigate complex systems which can change in numerous ways. Humans cannot adapt to conditions like this, so they are more susceptible to bugs or other failures.

While robots can be programmed with specific tasks in mind, people have a variety of goals. Human beings are seldom expected to operate on their own. It is almost always a computer that is required to complete a task. It is possible for humans to do tasks quicker than computers if the parameters have been set.

If you are trying to understand someone's meaning in a foreign language, or a difficult one, you will need to know the meaning of each word and find the right answer. It takes more effort than a computer to do the same thing. A machine has all of the characteristics of a human, which means it can work faster.

Robots, however have more in common with humans. It would not be the same task if robots were used, but there would be some error. What we mean by "error" is that there would exist a gap between task and output. This is the difference among robotics or automation.

However, even then it's not perfect. Non-robotic automation is one example. They automate the car processes of many people. The car itself cannot be automated but the man must still be able to decide and make an impact.

Robotics' advantages

Robotic systems should not be thought of as robot versions or copies of humans. Instead, robots should work more like people. The robot must behave like a normal human. But it shouldn't be capable of doing more than a human. It's the distinction between playing the piano or playing the pilot's piano.

Let's say you're learning how to play piano. The autopilot will then start to play if you stop playing the keyboard. However, it is possible you didn't do enough to make the autopilot perform at an elite level. That's what an autopilot would love to do.

Robotisation results in higher precision and less error which ultimately leads to a reduction or elimination of errors. Robots can accomplish their jobs more quickly than humans, which results in lower costs and increased productivity.

Robotic disadvantages

Robots will always be faced with disadvantages as they are robots. It is easy to see how robots interact with humans if you place a human in a position similar that of a robot.

Imagine a robot attempting to complete the following task. A foreign person would receive a package and have the robot use the eyes and hands like a human to deliver it. This would be the best scenario for humans, but the system couldn't do a great job using its human skills.

Consider, for example that the robot could use face recognition to identify the person receiving the package. The problem is that it can't recognize the person with its eyes and hands. A solution is to give someone a package that would make him feel the best.

Even though the robot can do most things correctly, there are still limitations. The robot may not be able to do all that humans can do. The robot does not have the greatest potential.

Robots will eventually become like human beings. As such, it will be limited. It will also need adjustments to allow robots as much autonomy possible in order to solve any task. We are smart people, so if we allow robots too much autonomy we will become confused and lose our bears.

Another limitation is the fact that robots are equally as competent as their creators. Programs can be created by anyone who is more creative than the average programmer. But, if you don't have programming experience, you might end up with robots that can only perform routine and repeatable tasks.

The conclusion?

Although robots can do nearly all of the work that humans can do, there are some limitations. They cannot do tasks that people would not like to do, are restricted in some ways and require judgement.

This limitation will make the system more intelligent for programmers but also force programmers into a deeper understanding the system. This will take more computing power, but it will ultimately be worthwhile as robots can't do certain things that humans are able to.

Chapter 10: What kinda engineering is robotics and what are its implications?

First, you need to know that there are approximately 80 million neurons in the human brain. A processor chip has a million. Robotics is all aimed at improving neuronal performance of the brain in order to perform tasks in realtime.

Animal

This is a very basic concept. In 1984, the first product that introduced a computer concept to silicon was the IBM PC version. However, your brain's true value lies in the huge number of neurons that can be used and the real-time processing they perform.

The biggest challenge is the amount of computer power needed to do the job. This is an obvious point when we think about AI and Machine Learning. However, when we talk about robots learning for themselves, it adds an entirely new dimension. This allows robots to learn through their own experiences and not depend on any traditional programming.

Movidius focuses on this: bringing Myriad 2 Vision Prozessing Units, or VPUs (VPUs), to market as an innovative way to create artificial Intelligence.

What are the pros and cons of microprocessors

"Many of our machinery components are quite static." The machinery drives itself by large amounts of data, and the production processes can use this data in order to make decisions about chips that change the future. "

What is the role for machine learning in robotics and other areas?

While computers are amazing today, regardless of whether they're tablets, smartphones or PCs - AI is what makes them work. While there have been amazing advances in the field of computing power and AI is easier to use, what really allowed this to happen was the ability to process enormous amounts of information.

The vision processing units, or VPUs as they are called, were designed with the ultimate goal to allow machines to learn through their experience and not rely solely on traditional programming.

He explained, "Many of our machinery components that you see in our everyday life are quite static." "The data is what motivates machinery and production processes. The data can then be applied to make world-changing decisions at a chip level.

Finally, our aim is to take our Movidius experience and expand it into many areas and develop solutions that extend beyond personal computing.

What are the difficulties in creating an AI platform

"Machine instruction is very data-intensive. Data can be difficult to work with."

Machine learning can be extremely data-intensive. It is also difficult to get access to data. Google and Facebook continue to update their data storage from large data centres, making it difficult for future predictions.

This new approach to development has been developed by us. It looks at the right data and creates the most intelligent machine. In some ways, our approach is similar to Google's: to find out which types of decisions the system can take to make it more intelligent.

How do different layers technology work together, such as the vision chip or the cloud?

There are three main parts to the world, the front, which is the display and the sensors and camera on the device. Your devices have the back-end, and you have the processing power they can run. It will be a chipset at the core. These chips aren't going to disappear.

We have had to work closely as we brought new learning machines into the market. AI's hardware continues to develop with the passing of time. While it is clear that the hardware's limits, which are located around the graph processing units, have disintegrated we also see that there are still machines available.

How do I know if my AI vision chips can be used in real life?

We are focused on making existing applications compatible with our vision processing unit. We are not happy with the functionality in a brand new application we have never used before. We are therefore currently working to develop products that are specific to the branch.

We also see ourselves more as a platform to help other companies and developers use the neural networks innovative application.

What is the future potential of the chip industry

Data and where the next generation is coming from are two of the biggest issues facing the industry. That's where our team gets into it. The cloud is the future for computing. We will find it easier to work with this data when it is transferred from devices into the cloud. Processing data is simple for a processor.

There is much computational work at the core of cloud computing. It's crucial to look at it from the business perspective. You will need the resources needed to back it up.

Nvidia once had you running the AI supercomputing group. What is the secret to this?

We are currently working on applications that employ the NVDLA processor. We've been in space since a while but it's much more than just a supercomputer. We are now working with companies and looking at other applications.

How would AI chip technology compare to other GPU architectures already on the market?

NVDLA processes large amounts of data. You will need a lot of computing power in order to make the right choices. You need to ensure that the data processed quickly and does not lose its accuracy.

It is the world's first chip that has some capacity. This also applies to the TensorRT platform which we are creating. It has a great potential.

How will this change the way Microsoft, Google, Amazon, Facebook use deep learning? Nvidia TensorRT architecture is being used?

We have created a highly-performant learning environment. Although we currently work with IBM Google, Facebook, and many other companies, we intend to continue to offer high quality service to all of our clients.

The GPU companies that offer these environments all work together. There is great motivation for all companies to use a common platform.

Is there any limit to the potential for learning technology that is profound?

The machine learning chips can solve many great problems. Some of the more difficult problems you can solve using deep learning require high-level image processing.

The next step will be data. Many of those problems that will be solved using the profound learning process won't be solved within five years. They will be solved in ten years. Now the question is: What infrastructure is required for that environment?

What is the future of education?

I don't really know. We don't have enough data so I don't feel confident. Look at what will happen autonomously. There will be many cameras mounted to the vehicle. Therefore, we need a lot o bandwidth to enable us drive.

I don't know what we will do, but I know someone will find a way.

Some of major technological companies are focusing on different segments of market.

You can take an example: The cars that drive themselves are going to change. It will be completely different.

What is frc-robotics?

Each couple of months we create a schedule, where people can travel, fight robots, and drink the beer that Robots have made in the past few days. Robotics have to fight robots from the side, not directly. Robotics allow 3D Robots to fight in a hostile environment and have beer.

What does robotics mean for healthcare? How can robots become doctors' substitutes?

Robotics (or robotics) is an emerging field that aims to improve the capabilities, processes and efficiency of healthcare providers and doctors. Robotics is an approach to healthcare that places patients at the heart of it.

How is geometry used for robotics?

We'll show you how these sensors work and give you some examples of robotic uses.

I recently attended a robotics meeting and was impressed by the number of experienced robotics investigators on the roster. It was fascinating to discuss what we did and the potential for what we did.

Many great articles have been written on robotics by individuals outside of the field. However, most focus on consumer robotics. I wanted to take a different view and examine some examples of robots in use in industry.

Robotics for design and manufacturing: For the production, design, assembly

A production background is essential to see assembly lines work like clockwork. Or that computer-driven machine tools work fast, accurately and efficiently. These machines are not magic but can be used to perform complicated tasks.

Let's first look at production robotic vision.

Since long, the automotive industry has been heavily robot-dependent.

What are the differences between artificial intelligence and robotics?

Were you aware of the fact that the most important and disruptive elements in the industrial manufacturing market have always been cognitive computing, robotics, software, and e solutions? Your competitors have already identified the disruptive elements and are aggressively using robotics automation and artificial Intelligence.

General Electric's partnership with Johnson & Johnson is not the first. Robotics alone is currently valued at $34 billion.

This niche within the IT industry has made our customers stand out for more than a ten year.

In 2019, and beyond, more of this will be possible. These tools will need to be developed faster than ever. What will be the biggest application of industrial automation in 2019?

Cognitive computing will play a major role in industry business models. This is because it assists companies to better manage business risks and create and implement new regulatory compliance risk mitigation and regulatory strategies.

How can we use the Internet of Industrial Things in 2019 (IIOT).

MGM Resorts International: Mr. Dan Singer; Global CIO

IoT can be used to gather, analyze, and enhance digital business models. This will allow organizations to deliver new products and services.

Big data and technology are going to have an ever-growing impact on how businesses and industries operate. Operationalizing Big Data is a continuous effort that requires more technology than simply adopting new technology. This requires a new mindset that links our business strategy to our technologies.

Industrial IoT is evolving from a simple collection platform to one that can manage the data. GE Predix 2018. Improved machine health management. Operators now have a single view, giving them a "heads up" of all the machines. These data collection and management tools.

Predix's 2018 release will mark a significant milestone for GE Predix. The algorithms embedded in Edge will be used to understand the data and send it to the cloud to make the required changes to GE Systems.

Companies will be able modify maintenance schedules based upon sensor data. Predix 2018 will be a key tool that operators will use to cut operating costs.

PTC Predicts IoT Future

PTC forecasts include:

Digital Twins will become the Internet of Things' (IoT) and Artificial Intelligence's (AI) leaders in 2019. We expect to see new business models and application creations within a short period of time and support new types of Digital Twins. Digital Twins (dynamic representations of a biological, digital, or physical system) will be created. These twins will then be used for IoT high-level projects and experiments. These digital twins are valuable because they can be used to reduce risk, increase efficacy and enable predictive maintenance. The digital twins, which use the new technology for infrastructure and data centers scenarios, will allow for the creation of new IoT application types.

While digital twins have been developed for a specific type of case, there are now digital twins designed for many different processing processes. A wide variety of digital technology options are available, which is a benefit of using digital twins in process applications.

The digital twins are expected to enable new industrial scenarios. This should happen by the end 2019. These scenarios will emerge from the emerging physical processes. This feedback loop between the digital-physical worlds will drive the emergence and development of the new digital twins.

Industrial IoT's future success will hinge on cloud computing. SDN is a third-generation Internet protocol, also known as IP 3.0. This allows companies to continue to make use of advanced IIoT techniques.

The cloud adoption will accelerate as more companies adopt SDN and cloud. IoT firms and other companies won't adapt will lose ground to these important trends.

SDN and cloud will become a precondition for the adoption of IIoT technologies by many businesses. Their IIoT solutions will attract more companies to host their IIoT services in the clouds and make them available to partners and customers. AWS and Azure offer the easiest path to cloud and are most likely to scale.

The future will bring significant technology advancements and make it easier to access the next generation IIoT apps. Digital twins that can be built quickly and deployed in a matter of weeks will become a reality. This is a significant technological advance over the past few years. Digital twins have digital sensors that replicate what's happening in the natural world. These twins will be capable of reproducing what's happening in the actual world.

In the coming years, however we expect there to be a slight slowdown in market activity. IoT is expected to continue in a mature market with cloud maturity. This will allow firms to make informed decisions about IIoT potential, technology maturity and priority area areas.

What is a high school robots team?

LANSING, Mich. (WLNS). - Robotic teams at high school gain national attention.

There are also FIRST Robotics Competitions where high school teams compete against other teams.

More than 600,000. students will be competing in more then 30,000 teams.

Nicole Lampa (6 news reporter) spoke with Matt Henry (student at his school Grand Rapids West Catholic and one the FIRST Tech Challenge Team members of 1714).

I spoke to Matt Henry and his colleagues as part of FIRST Tech Challenge.

They are all high school seniors. The team they make up is called Team 1714.

They spent most days building this robot and at school.

I asked them their team names.

Matthew: "Team1714" is not a popular name. Everyone is confused by the name 1714. Different reasons exist, but the one thing that unites us is our willingness to go above and beyond what is expected of our team.

Brandon: "Well, my name would be our name. We are making a return, so it is reinvention. It is like getting back to where we came from. Over the last two year, we have rebuilt our team.

I also asked the group why they had decided to create their robot.

Matt: We have to create a robot that does as many tasks as possible in order for us to win this competition.

In fact we have changed how we do competitions this year.

We do not have robots for team building, so we use them to make sure people get along.

Robots that can perform all tasks

Like a robot supervisor.

Everyone involved in the contest doesn't know what their task is until the robot completes it. Also, it is important that we design our robots together so that the robot supervisor can work with them.

The FIRST Tech Challenge was a success and the robot was awarded the chance to participate in the state competition.

Matt: "We were just as excited to be there again and that we didn't need to deal with any pressure that we could choose which robot to use, that was all we had to do.

Robotics is a growing trend in Lansing. Michigan is also known for hosting various competitions.

High-tech robotics & optics

ROOSEVELT ISLAND Papua New Guinea - This small bird, about the size of a potato, is a marine bird.

He's also a seafarer experts -- a "sea robin" can drill food under the ocean surface. The depth can drop to 35m, twice that of the military's deepest submersible diving.

Remotely controlled submarine vessels and robots have transformed ocean science.

Tim Shank, University of Queensland professor says, "Robots will change the nature of science research." Shank is the leader of a project using sea robots to map and analyze the seafloor. He is however adapting existing robotic technology, rather than creating it.

Shank, CBS News: "We can use drones. Flat wing drones. We can then use underwater vehlcles. To explore the bulk, we can use the seafloor."

"Imagine, for example, a robot capable of using instruments in the water or submarine mapping."
"

Navy SEALs are already researching robotic systems for submarines. These technologies aren't cheap and the U.S. Navy isn't going to stop relying upon the same mission.

Shank stated that robots are a legitimate part of developing the next generation in technology.

Shank and the team have built a small underground vehicle, which can rise to the surface of the sea and map its environment using different sensors. This vehicle can cost up to $1,000,000, but it will power next-generation ocean exploration technology. He explained that they are much cheaper to maintain, have lower operating costs and provide more power.

The challenge is to find a cheaper and more convenient way to propel these underwater vehicles. Shank said, "You have to make our robots as simple for people to operate as possible."

X-47B Unmanned Combat Air System U.S Navy

He also said that the Navy's focus is on the reliability as well as the complexity of robots.

Shank stated that "it's not just about the type of robot, it also concerns both the strength & robustness the controls and also the strength & robustness the guiding / control algorithms."

The Navy is still working on designs but some ocean robots can be seen being tested.

News of CBS

The U.S. Navy's X-47B Unmanned Combat Air System played a significant role in initiating the first U.S. Navy manned flight towards orbit and then to a vertical land and recovery.

Shank stated that after a self-sustaining recovery has been a success, the establishment will think the second autonomous recovery is safe.

The U.S. Navy has been testing a remote vehicle for deployment in 2020.

Navy and Shank also explore autonomous submarine technology.

He stated, "They are completely independent." "This will give them the ability search for autonomous target.

This could be a three-man crew, but Shank suggests that at least one crew member remain aboard.

Shank stated that it was necessary to have someone present at the end, as it is unsafe to send a machine to hurt.

How great is a robotic Hanson!

Scientists and educators primarily focus on cognitive tasks. Learning by doing is key to making people think like robots.

People can learn to do a certain task in a way that imitates real-world learning, which is much more effective than teaching computers. An example of this is a robot that can drive a vehicle or slice human flesh. They are now closer to their consciousness and no human guidance.

But is it possible to teach a robot the same learning techniques? Is this an apprenticeship that is conscious of itself, or based on its individual interests?

You can achieve this by designing robotics that can learn and adapt to changing situations. A robot in a lab may choose to do some tasks when the reward will be greater.

Another approach is to build a robot from scratch that learns from its experience. This robot could be used to drive a car or other vehicle. It starts with driving a vehicle and then it explores different driving situations.

The first step in human robot training has been effective in some ways. Robots are programmed to learn from trial and errors and to motivate themselves. They can make random errors to help improve their work.

Yet, it is important to learn from experience because robots can identify certain causes and implications so that their decisions are better based on what they see.

Although it may not appear to be a sensitive choice, pressing the selfdestruct button could prove to be a sensible one. It could, however, be discovered that a robot can observe how people react to a simulated collision and may help prevent one by pressing the selfdestruct button.

And now, the notion of "emotion", also known to be called "emotionally smart robots"

Robots, Emotions

Robots' concept of emotion is controversial. It isn't clear what emotion is, and what robots could display it. Sometimes, the definitions of emotion are different than what people might consider emotional behavior.

There are arguments that robots do not have emotions, or that robots can be programmed emotionally.

However, some people believe certain behaviors (e.g., laughter) to be more emotionally aware.

It is possible to create robot systems that understand your environment and adapt to it by using algorithms that match what you do.

If a robot wishes to learn how it can perform a particular task, it may be programmed to work with other robots.

Once a robot successfully completes a task, it may decide to share its performance with a group of robots. This feedback can help the robotic learn what works and what doesn't so it can continue to be successful.

Is it possible for robots to be sufficiently intelligent to make such decisions based solely on their own experience?

For now, it's useful to compare human behavior and that of robots to understand what could be considered emotional.

Robots that are able to read the emotions of people after interaction could be able learn what makes them happy and what makes them angry.

What Is Robotics and How Can It Help You?

Robotics is a merging of diverse scientific areas. This mainly uses advances from the following fields.

* Manufacturing technology

* Material science

* Mechanical engineering

* Advanced algorithms

* Fabrication sensor components

Anyone who dabbles in robotics will discover that they are interested in many types and methods of scientific practice. Many people begin robotics because of a romantic idea about the benefits that robots have for humanity and the world. Imagine how the world might change after the successful implementation these remarkable machines. Images like those seen in science fiction and movies are amazing. If robotics interests you, it's time to start exploring the world of robotics.

Robotics is, as we have discussed briefly, the study and design of robots. That's only a brief explanation. Let's dig in deeper. What makes a bot, a bot? This topic can quickly become a philosophical issue, technical discussion, or both. It is best that you learn the most common definitions first. Once you have a good understanding of the basics, it will be easier to move onto more complicated topics.

Mechatronic Devices

Robotics is not the same as robot. A robot is a machine or thing that can carry out any type action or behavior. Robotics is both the theoretical process and the practical application of robots. This could be your garage door that opens when you approach. The door would have a sensor to detect any remote signal. It would use its actuators in order to physically open the doors. Once it detected the right signal, it would switch off its motors after it closed the door. Although technically it is robot, the term "mechatronic system" is better. It is still usable for our purposes, however, and we will be introducing the concept of robots.

The main components of a mechanical device are:

* Sensors for detecting what is happening around your device.

* Actuators, to move and do real work.

* Control system, which allows you to operate the actuators without being aware of their impact on the environment. This is done using the sensors and guides.

Robots with "True" Value

It is possible that the mechatronic gadget we talked about above is not a true robot. The components are the same, but the device also needs additional parts. This is covered in detail under the Key Components chapter. To be considered a robot, the device needs autonomy. It should be able do the tasks by itself without the aid of someone. Can you see the basic inabilities of the "automated" garage door we showed? To activate the door's opening, you would need to press a button on your remote. It is not a robot.

Many people will disagree on what constitutes a robot. Many would claim that the garage door can open on its own, even if it is only when a human signals it to. But it's not that important right now.

These two robots are extremely common.

* Automated machinery, which is a machine that has its own actuators and sensors to tell it how to move. They can accomplish any task without human intervention. This could be as simple a vacuum robot. This type is very popular in households. It could be a cleaner. The vast majority of industrial and commercial robots will be classified under automated machines (or autonomous devices), which has the exact same meaning.

* Pet machines that are capable of moving at least a little. They can sense what is going on around them and make decisions based on that information. They can also sense what is happening around them and take action. These robots can be used to entertain people and help their creators understand robotics. Making a robot pet is an excellent way to learn robotics in fun ways.

You've likely seen machines that look like robots but are not robots. This includes fighting robots, which are extremely popular. They can be used to teach you how to design and build many of the components needed for robotics. But they are not robots. They're remote-controlled machines much like radio-controlled cars or planes you might have owned as children. They can be very heavy hardware, but they lack autonomy or the sensors to be mechatronic.

If a machine is capable of performing its tasks autonomously, it is usually a robot. Other than that, it could be very close but it's not robot.

Robotics in History

Let's take a look at what a robot actually is. It's a machine with the ability to be programmed to perform at most one task. It does not necessarily have to look exactly like a person, but it can be programmed to perform at least one task. The idea has been around since its conception.

The term "Robot",

The actual term "robot", was first used in 1921. Karel Capeck (1890-1838) was a Czech playwright who is believed to have been the first to use the phrase. Although it's possible his brother was the one who suggested this term to him, Numerous times, he was named as a Nobel prize winner candidate. Because of his vast work as both a playwright, and an author, it was no surprise that he was considered a Nobel prize candidate.

Capek used it in his 1921 successful play R.U.R. Rossum's Universal Robots, a company which made robots, was the subject in the play. Due to the many marvellous things robots have done, the play depicts a utopian society. While robots provide happiness and a better lifestyle at first, eventually, unemployment becomes a serious problem. This sort of thing is common throughout history. As new technology is developed that can do the work, it's possible to see.

Early History

The idea of an automatic helper dates back to ancient mythology. Pygmalion is a tale about a statue coming to life. Cadmus is said to had planted the teeth from a Dragon, which later formed into soldiers. You can also find golems of clay in Jewish mythology. Norse legends included giants made of clay. One Chinese account around 1000 BC described automated things that looked almost human-like in a Chinese account.

As we can see from their literature, the idea that automatons existed was extremely popular during medieval times. Automatons became more common in the 1700s. People were often amazed by the complex designs that were created. It wasn't until later times that machines like the newer robotic ones, which we will focus on in this chapter, were invented.

Modern robots, which are electronic and have advanced programming, look very similar to their older cousins. They are not all that different from the first machines built by humans. You might be surprised to learn that although many of these contraptions were novelty items, they represent a great deal of thought and scientific application.

There were both humanoid automations and animals that could work. These types of machines used low-tech systems, such as weights and water, to generate power. A large industry was created for clockwork machine manufacture after medieval times. This included the use of springs and levels. These were essential for simple tasks. These devices are not modern by today's standards but they were quite ingenious back then. They provided a foundation for robotics scientists. That knowledge has enabled robots and machines to progress to where they are today.

Isaac Asimov's Three Laws of Robotics

It shouldn't surprise you to learn that science-fiction authors have had a great deal of influence over the core ideas behind real technology. Star Trek fans will be proud to say that many of the most influential inventors in the world have a deep love for Star Trek. Isaac Asimov (1919-1920), an American writer, was a success during the Golden Age. You might have read his story, I

Robot. In his story, he outlined three principles that govern how robots should behave. He later added another, the zero rule.

* Law one: A robot can't injure another person, or let a human being get hurt through inaction.

* Law Two. A robot must respect orders given to it by humans, except in cases where they would be inconsistent with the First Law.

* Law Three. A robot must maintain its own existence provided it doesn't violate the First Law or Second Law.

* Law Zero. * Law Zero. A robot can't injure anyone or cause harm by inaction.

These laws are often referred to or modified over time.

Unimate the First Robot

As you may already know, World War II led to a lot of technological advances. Although war is clearly not a good idea, the drive for success and the ability to beat your competitors has produced a lot in the way of machinery and new ideas. It was 1945 when George Devol, an entrepreneur and inventor who became a success, and Joseph

Engelberger (an engineer), began to talk about Isaac Asimov's writing.

The men decided to join forces to create a robot. They wanted to create a robot that was both commercially viable and functional. Condec Corporation CEO Norman Schafler was impressed by their idea and voted for them.

Engelberger began to create Unimation. This was a shorter version of the "universal automaton" and they were able to get the support that they needed. This was the first ever commercial company to manufacture robots. Devol submitted the patents needed to enable his partner to set up the manufacturing portion of the company. The robot they created was called "Unimate" after their company. Engelberger is also known as the "father and pioneer of robotics".

Unimate robots were first put to use at a General Motors manufacturing plant. Because die casting was something that was hated, the robot was placed to work. Unimate robots continued to do similar tasks and even spot weld cars at the factories. This was a big success commercially. Robots were easier and more cost-effective than people and were quickly adopted in other industries. The field of robotics is not just one that was scientific. Robotists and their customers had the potential to make some serious cash.

Robotics Application (Uses)

The chances are that you have an idea about what a robot might look like. Robots are typically used for jobs that are too dirty, dull, dangerous or dangerous for humans.

There are two types currently in use in the real world:

* Dedicated robots which can be used to accomplish a specific task or perform a limited range of tasks.

* General purpose autonomous robots which can do many different tasks.

Autonomous Robots For General Purpose

These devices are capable of performing a range of tasks. They can navigate in spaces they are familiar with, use doors and lifts to interact with other objects, recharge when they are short of power, and perform many other tasks. These are simple tasks that a human would consider trivial. General purpose robots also have the ability to link up to other networks, much in the same way a computer can, so that they can access more resources, and become more valuable, resources such as more software or additional computing power.

These are the tasks that general purpose autonomous robots often perform:

* Interacting with people using speech and motion

* Recognizing individuals

* Acting as a human company

* Monitoring the environment's quality

* Responding when an alarm goes off

* Collecting Supplies

* Other menial work

It might not seem that the list above is anything extraordinary for humans. The beauty of these robots is their ability to perform multiple tasks and switch between them, without needing any programming or refit. They can also spend a whole day going from one task into the next, much as a person would.

Some robots look and feel like humanoids. Even though they are still relatively primitive, these robots have been a part of the human imagination for a long period. Even in familiar environments, humanoid machines are still unable perform the above tasks without error.

Dedicated robots

These are the real robot workers currently in use. The robots used for general purposes (or humanoid) might not be as popular in the near future, but they are the ones that get the job done. Robots already do a large portion of the world's job. It might surprise you to see how many jobs robots are taking over. If you're concerned about such things, or anything else that has to do with robotics, then please read the Future of Robotics chapter.

Commercial Robot Uses

These are just two examples of robot workers currently performing different jobs:

* Electronics fabrication. Robots are nearly all responsible for the production of large quantities of printed circuit boards (PCBs). They are a pick-and place machine that can lift electronic parts off of small shelves and put them onto the circuit boards. These robotics are capable of doing this with extremely high accuracy. Their speed and output are unbeatable by human workers.

* Material handling. Robots all around the world are capable of handling manufactured products and their material. Robots can be used to move packaging material or to operate loading and unloading equipment. Around 38% percent of all robots are used in material handling. However, this number continues to grow.

* Dispensing. This kind of work involves painting, spraying or gluing, as well as sealing with adhesive. Although robots can perform precise, smooth jobs, dispensing tasks account for only 4%.

* Picking. Amazon is one company using robotic workers. They have small, box-looking robots that can actually transport shelves of items to workers.

* Assembly. In the assembly area, robots can do pressing, fitting, fixing or disassembling, as well as inserting. Robotic technology has allowed for a wider range of applications. This has led a decrease in robots used for assembly. These new technologies are tactile sensors and torque sensors, which provide superior sensory capabilities for the robots.

* Car manufacturing. In automobile production, robots are a common tool. A car factory has many robots. These robots form production lines that are fully automated. A total of ten robots can be found for every worker employed in a factory that makes cars. The robots are installed at different stations in the assembly line. They can complete entire sections independently, so there is no need to have human help.

* Welding. Robots can do most types of welding using robots, such as spot or arc welding. Similar to the first commercial robots, they are primarily used in the auto industry. Even smaller manufacturing factories are beginning to use robots for production. It's becoming more affordable to use robots as the prices for them come down.

* Processing. The tasks of processing robots include cutting with water or laser beams. This is a very small portion of the robot industry. Only 2% are used for processing. This is probably due to the ease of using simpler automated machinery for the same tasks.

* Military Use. The military has some of best robots in the universe. Many consider them to be the most important. They are the high point of robotic technology. Many people have come to depend on them for their safety and survival. It is important not to forget that robot use in military operations is a topic of heated ethical and political debate. The following are some of the most common robots used by the military

o EOD. Explosive Ordinance Device Disposal robots have the ability to detect and deactivate mines.

o UAVs. Unmanned Aerial Vehicles fly without the aid of humans. They are used for spying, scouting enemies and providing a wider view of battle area.

Robots: The Benefits

Robots offer many benefits to humans, their workers and the entire human race. Robots for commercial use could dramatically improve the quality life on planet Earth, if properly introduced. They could reduce the number of people working in dangerous, filthy, tedious, and hard jobs. Many people are concerned that the increase in unemployment could cause large-scale shortages and mass unemployment for the majority of the workforce. However, that's all still open to discussion.

Robots have the potential to improve the efficiency of industry by giving greater control over the management of their workforce and ensuring that products and services are of high quality. A robot can work throughout the day and not get tired. Apart from maintenance issues, robots are not affected by fatigue and wondering thoughts.

The use of robots to cut down on the cost of manufacturing can make it possible to lower the overall cost of items. This would bring immense benefits to all nations, especially those that can't produce enough basic necessities like food and water to provide for their citizens. But that's another story. The Future of Robotics chapter covers more of that.

Robotics Jobs

Are you interested to pursue a career working in robotics? You've probably noticed that not everyone working in this industry works directly with robotics. There are many areas that are connected and feed into each other to create new technology. They also impact other areas such as research, power production, and manufacturing. Robotics and related jobs are growing.

Robotics is now largely concerned with lifting and moving of objects. Robots are specifically designed to perform specific tasks at this stage in human history. A grown human being could simply walk over to a bag and grab it. Then, they would be able carry it to the desired destination. An enormous amount of analysis must be performed for every action a robot is expected to take.

Once all of the robot's smaller actions have been analyzed, you can put them into practice with programming and machinery. These tiny actions become a larger piece - the actual robot. It takes a lot more than one person to figure out how to do something simple like move a piece of furniture from one place into another.

There are several common jobs in robotics.

Operator

You might consider a job as an operator if your goal is to work directly with robots. These jobs comprise the majority the workforce of the robotics field. They include pilot technicians, who can be described as electromechanical technicians. They work with robot launch and recovery systems. They also use the tools that are attached for robots. Operators will assist robotic engineers in the design of new equipment and their improvement. Operators also will be able repair the tools they work with.

Robotics Technician

These workers are often supervising robotics engineers. They are responsible for constructing the systems that the engineers designed. They also have the ability to troubleshoot and fix any issues with the designs or maintain documentation. As well as helping with robotics design and construction, technicians might also be involved in their maintenance. Sometimes, they might be needed to incorporate ideas from engineers or design a platform for robotic machines to work on. They aid in installing robots by hooking up wires or putting actuators in place so they can function.

Robotics Engineer

Engineers need to have training in electrical engineering, computer science or electronic engineering. No matter their field of specialization, a robots engineer must be proficient in using computer programs to solve problems. They will have to know how to install robotics equipment. They will often be responsible to supervise workers who are building robotics systems. They inspect and review the designs of robotics technicians. They also test and fix any problems.

Similar Jobs

These are some examples of indirect jobs associated with robotics.

* Installation safety system

* Welding

* Repairing or maintaining robots

* Operation of hydraulic equipment for testing

* Failure analysis testing

* Programming/reprogramming robots

* Interpreting schematics

* Installing or removing robot equipment

On average, a robotics technician would earn $55,000 USD in December 2014. As a median salary robotics engineers made $75,000 USD

Key Components

There are many different types of robots and they're all used in many different situations. They have some basic similarities, even though they can be used for many different purposes and industries. This was briefly explained in the What Is Robotics? chapter. The book's opening chapter covers this topic. We will now explore the inner workings of a bot in more detail.

These basic components are all found in modern robots. But don't forget that there are thousands more parts that can be put into just one component.

Three Key Aspects to a Robot

Form

A robot will need some kind of physical form. This may be a form of frame, or another shape that allows the robot to do a certain task. An appropriate wheel may be used for a robot that is intended to travel on a particular road. Depending upon the intended function of a robot, the type of form used may vary. Robotic engineers create the form they need to do a specific task. This can lead to some rather bizarre robots. There are plenty of boring robots available, however, that have been made to be useful and practical.

Electrical Components

A robot can only do what it is supposed to do if it has electrical components. These components control the robot's machinery and provide power. The previous example of a robot that needs power to turn its wheels would be the one in need. This will most likely be in form of electricity.

This will need to travel through wires on the robot and from a battery which stores electricity.

Even if the machine is powered by petrol or diesel, it will still require electricity to start a engine. This is why engines and cars need batteries. Additional electrical features of a robotic device include motors for movement and sensors for monitoring the surroundings.

Computer Code

In order to "think", robots must be programmed in computer code. These robots would not be capable of deciding what to do next or how to act. In other words, they would not be robots true to the meaning of the term, but just mechatronic gadgets. On the contrary, many non-robots still have computer code.

As you can see, a well-constructed form and power to enable movement and sensory capabilities are essential for a wheeled bot. It wouldn't be capable of moving automatically without computer programming code.

Programming powers robots. Without a great program, even the best robot design may not work as it should.

Here are the three types and descriptions of robotic programs.

* Remote control. This is different to a traditional control that is controlled by a user. Remote control programming is made up of commands that can be programmed ahead of time. The signal will be received by the robot and it will then carry out the programmed task. A human usually gives the signal, which allows the robot to do other things. These robots can be called automations. This is not the purpose of this book.

* Artificial intelligence. AI robots in the truest sense of the term are what many people associate with. This is largely due to the numerous movies and books that have been made over the years. An artificial intelligence program allows a robot to interact with everything around it, without the need for any signals from any control source. They can respond to objects and occurrences using the programming code that tells them what to do.

* Hybrid. This is a hybrid of remote control, artificial intelligence programming, and remote control.

Further Components

Sensors

These allow a robot learn about the world around them. These sensors allow a robot monitor what's happening to their own internal parts. Sensors are required to enable a robot to perform any task. Sensors allow a robot detect changes and to take the necessary actions. In our earlier example, a robot is required to drive down a roadway on wheels. It must be capable to detect any changes In the road or obstacles.

Sensors are used in safety features and to alert robot operators when there is a problem. This gives the robot real-time information, which allows it to see exactly where it is located and what it does.

Two of the main sensor types in development at the moment are vision and touch. Sonar and radar are two other types that are common.

Actuators

These can be described as the robot's "muscles". They are the part that converts stored energy into motion. By far, the most widely used type is an electric motor. They turn a gear or wheel in order to create movement (as you do with your car). Linear actuators can also be used, but primarily in factories as robots.

You can choose from:

* Electric motors. The majority of mobile robots use DC motors. However, industrial robots require AC motors. AC motors have the ability to remain stationary and continue their work. These motors work best when some form of rotation is needed and when very heavy loads are not being carried.

* Linear actuators. These are available in a variety of forms. They can quickly change their direction and go in and out. They can move heavy loads easily and are often moved by compressed air or oil.

* Shape memory alloy. This is also known to be "musclewire". It can contract when it is powered using electricity. However, once the electricity is removed, it returns to its original form. Because of its simple and lightweight design, musclewire is ideal for small robotic components.

* EAPs. These are electroactive Polymers. A type of plastic that is capable to contract up to 380%. It functions in the same way as "musclewire", but with electricity. EAPs are used in the arms, faces and hands of robots made to look like people. EAPs can be used to give robots the capability to walk or fly, float, and swim.

* Piezo motors. These are an alternative to DC motors. They work differently and can be used for different applications. To create thousands upon thousands of vibrations every second, tiny ceramic elements are used. This generates either linear or rotary motion. Piezo motors can be used for a variety of purposes, such as vibrating a nuts, screwing in a screws or moving a vehicle in a common way. Why would someone need to make something so complicated when electric motors can be found. Piezo motors provide tremendous force and speed, regardless of how small they can be.

* Pneumatic artificial muscular. These are also known to be "air muscle". They are expandable tubes which can be moved using air.

* Series elastic actuators. These springs are combined with a motoractuator.

* Elastic microtubes. These actuators have great potential, even though they are still in development. Carbon nanotubes can store huge amounts of energy as they are free from defects. One 8mm strand could do the exact same job as a human's biicep. This type "muscle" power could help robots become even more powerful than people in the future.

Power Sources

As you probably know, batteries are the main power source of robots. There are many kinds of batteries that can charge robots. When considering power supplies for robotics engineers, there are three main factors to be aware of: the battery's safety, life span, and weight. As a traditional engine, a generator can also be used.

A robot can be tied to an external source of power without the use of a battery. This allows robots that are smaller and lighter to function, but the robot cannot move completely without a cable.

These power sources could also be used to power robotics.

* Solar power

* Nuclear

* Pneumatics

Flywheels are useful for storing energy

* Liquid hydraulics

* Organic waste including garbage and even feces

Manipulators

Robots must be capable of physically manipulating things such as picking up and altering them or even destroying their contents. Manipulators include the robot's "hands", also known as the end effectors, and an arm (also known as the manipulators).

A robot's effectors, or hands, can usually be replaced. This allows them to be reused quickly for different tasks, without needing a whole new piece. This is like someone who puts down a screwdriver to be able to grab a hammer.

There are robots equipped with fixed manipulators that can't move. Some robots can also be equipped with special manipulators that can be used with various effectors.

* General purpose. This effector looks most like a person's hand. They are usually used with the more advanced robots which are often humanoids. General purpose hands are extremely dexterous, with hundreds of sensors providing tactile feedback to the robotic arm.

* Grippers. The most commonly used effector type in robotics, are mechanical grippers. The simplest form of gripper is comprised of two "fingers", each of which can be opened or closed. The robot can then pick up or release things by doing this. Some mechanical grippers mimic the human hand while some are more machine-like in appearance.

* Vacuum. These can be used to hold heavy loads. However, a vacuum grasper can only manipulate large objects with a smooth surface. Vacuum grippers often come with robots that can pick up and drop things like windows.